MANAGEMENT
OF ENVIRONMENTAL
IMPACTS

MANAGEMENT
OF ENVIRONMENTAL
IMPACTS

ALANDRA KAHL

MOMENTUM PRESS
ENGINEERING

First published in 2018 by
Momentum Press®, LLC
222 East 46th Street, New York, NY 10017
www.momentumpress.net

ISBN-13: 978-1-94664-691-0 (print)
ISBN-13: 978-1-94664-692-7 (e-book)

Momentum Press Environmental Engineering Collection

Cover and interior design by S4Carlisle Publishing Service Ltd.
Chennai, India

10 9 8 7 6 5 4 3 2 1

Printed in the United States of America

ABSTRACT

The various types of environmental impacts of projects are discussed in sufficient detail as to provide a concise and useful overview for the graduate or professional student. Subjects approached herein include impacts on water bodies, generation of hazardous materials, and alternative energy sources and their impacts, and management of sustainable projects. The intent of this volume is to provide a repository of general information for consultation and reference of the user.

KEYWORDS

design, environment, environmental impacts, life cycle, project management, risk, sustainability

CONTENTS

LIST OF FIGURES

ACKNOWLEDGMENTS

Dr. Kahl would like to acknowledge the support of friends and family during the writing of this book.

CHAPTER 1

INTRODUCTION

The management of environmental impacts is a topic that is critical to the health of the greater environment as well as relevant to science, engineering, and industry. To best ensure a sustainable world for the future, the impacts of industry on the environment need to be effectively managed and assessed. Most major industries have an assessment framework in place that aims to assess and reduce their impact on the environment. This framework frequently includes an environmental impact assessment plan and individuals devoted to the compliance and management of environmental regulations.

The assessment of environmental impact entails evaluation of potential harm to the environment from a project as well as the repair of previous impacts from earlier projects. Impact statements may address one or both of these aspects. Assessment of environmental impacts is done with regard to impacts of all aspects of the environment, including species harm, land, water, and air impacts. Each of these aspects must be enumerated fully for an environmental impact statement to be complete.

Environmental impact studies ensure that developers take responsibility for the safe protection of the environment during their proposed actions as well as those actions that may take place in the future by the same individuals on the same site. They also require that any impacts that may occur during the lifetime of the project are properly managed and mitigated. Environmental issues that arise during a project review may include managements of waste products, habitat impacts, or land and air impacts, such as particulate pollution.

Assessment and management of environmental impacts inform regulators and developers about the viability and longevity of a proposed development as well as whether that project should proceed. In order to move forward, a project must effectively balance the health of the environment with the growth of the industry. This includes compliance with and management of local, state and federal environmental regulation.

Effective monitoring of environmental conditions is a key component of the management of environmental impacts. Monitoring can be on a small scale for an individual site, or a large scale for a facility. Examples of monitored conditions include water quality metrics, soil analysis, and air quality. Data from these studies inform decisions related to project outcomes and goals. Environmental monitoring is also an important aspect of permitting and regulatory compliance, which are important to the effective management strategy for the environmental impacts of a project.

Community and stakeholder views are also an important part of the management of environmental impacts as these groups inform and drive the management process. Stakeholders include both the developers and the local community which are impacted by the development of a project. It is important to engage all groups in the management and assessment of the long-term vision of a project to best inform environmental decisions. The evolution of the environmental impacts from a project can also be driven by stakeholder views as these views shape the regulatory framework and can determine the proper pathway for management and assessment.

CHAPTER 2

ECOLOGICAL IMPACTS OF PROJECTS

2.1 CONSTRUCTION IMPACTS

2.1.1 PROJECTS ON LAND

Construction projects can cause significant disruption to the environment and ecology of a site, if not managed properly. Land impact can vary from small impacts such as blowing dust to large impacts such as habitat or ecological community loss. All impacts should be properly studied prior to the commencement of a project so that a proper management plan can be written.

One of the largest areas of construction impacts on land is that of road building. The United States contains over 4 million miles of roadways which take up an estimated 20% of the available land area. Roads are important to new development as they provide access to areas that were previously unable to be accessed by vehicles. Road building is a significant area of growth and impacts, with an estimated 60,000 roads to be constructed by the U.S. Forest Service alone in the next 50 years (US EPA 1994).

Impacts of road building are varied. For mammals, building a road can affect travel, breeding, and foraging habits as well as increase animal mortality. In fact, road mortality is the leading cause of animal mortality on roads in the United States, with an estimated 1 million vertebrate animals killed by drivers each year (Noss 2002). Some populations can be significantly impacted by roads if they pass through pristine habitat. For example, when interstate 75 was completed through a major deer wintering area in northern Michigan, deer mortality increased by 500% (Noss 2002). Vertebrate populations are the most visibly affected fauna by road construction; however, invertebrates are also affected by being run over or smashed onto windshields. These invertebrate populations can also increase vertebrate mortality, however, as predators follow prey into roadways. For example, the chimney swift follows insects close to the ground. When these birds follow prey over roads, it increases their chances of being struck by a vehicle. Natural defense mechanisms also cause animal deaths. For example, turtles perceive cars as a threat and draw into their shells to protect themselves, which increases the time they take for them to leave the roadway. On highway 27 in Florida, which passes over a lake that is important turtle habitat, the rate of mortality is very high due to this defense mechanism. Snakes also become immobilized when threatened and will often rely on their venom rather than movement for defense. It may take up to a minute for a frightened snake to leave a roadway after a car passes.

Roads are attractive to wildlife, which also leads to increases in their mortality rates. Reptiles and other ectotherms gravitate toward sun-warmed roadways to bask. Unfortunately, the numbers of snakes and other reptiles killed on roads are not well tallied. Herpetologists have made some estimates, however. On the highway passing through Paynes Prairie State Preserve, an area known for its diversity of snakes, herpetologists noted an important decline in the volume and variety of species when the four-lane highway opened (Field et al. 2016). Mortality of other species increases due to attractive behavior as well. For example, birds go to roadways to retrieve pebbles and seeds to aid in digestion, whereas other larger mammals are attracted to road salt and deicing fluids. Browsing herbivores such as deer often find attractive forage next to roadways due to tree clearing, and small mammals seek roadsides for habitat use.

Road mortality rates are also linked with migration and movement patterns. Animals may move to new habitat to mate or seek diverse sources of nutrition which can increase their mortality if a road must be crossed. Hatchings of large populations of animals such as reptiles will increase in mortality if near a roadway as the young disperse when leaving the nest. Animals with large ranges will also be impacted by roadways as they must inevitably cross a roadway at some point in their roaming. An example of an animal with large range would be the Florida panther, whose range has been measured to up to 630 km^2. Large mammals often have greater forage ranges, making them vulnerable to becoming roadkill.

Species fragmentation due to migration is also an important impact on the overall animal populations. Migrating female turtles for instance can affect species balances as the numbers of the females decrease while the males who do not migrate remain steady. These demographic shifts can directly exacerbate population declines and threaten population viability. Populations with unequal demographics are vulnerable to genetic deterioration and stochasticity (variation in age and sex ratios).

Habitat fragmentation is another important side effect of roads. Some individuals will not cross roads, resulting in smaller populations and fragmentation of species. Population isolations can reduce species variability and viability. Some animals such as snakes have shown documented avoidance behaviors of roads, which will limit their range of predation. This can lead to over-predation of an area and proliferation of a single species that can unbalance an ecosystem. Birds that typically fly short distances from tree to tree may be hesitant to cross the large open space of a road resulting in longitudinal rather than lateral populations. When species are isolated by roads, gene flow is restricted. In a study of timber rattlers, genetic diversity was found to be lower in isolated populations when compared to contiguous populations (Didham 2010).

Habitat fragmentation can be especially impactful when considering the effects of climate change. Populations that cannot or will not cross roads cannot adapt to the effects of climate change, such as reaching a water source on the opposite side of a roadway when a local source dries up, resulting in extinction or severe decline. Limited genetic variation due to habitat fragmentation can also prevent animals from adapting rapidly enough to prevent extinction due to climate change. This type of decline and fragmentation can also extent to plant populations that depends on animals for dispersal. Animals that are isolated by roadways cannot disperse seeds beyond their home range, resulting in lower plant diversity near the roadway as well.

Another land impact during construction is pollution. This type of ecological impact can range from an auditory impact (noise from construction equipment) to physical impact such as grading of the land. Noise in construction project persists after the project is complete as well. Vehicle traffic along new roads or new parking lots, industrial sounds from rail line or conveyors, all contribute to noise pollution on the land. Auditory pollution can be detrimental to humans if exceeding EPA levels, and can also impact local wildlife. Birds may find it difficult to mate if their calls cannot be heard, and other predators that use sound to locate prey may die off as well. Pollution from land project can also take the form of air pollution. Vapors from construction vehicles may be harmful to the environment as they degrade air quality by producing greater quantities of carbon monoxides and smog. Cars on roadways give off heavy metals from oxidation during combustion processes. Lead from gasoline persists for long periods in soils and may even now be found near highways. It also has long-term toxic effects, increasing mortality rates and disrupting reproductive cycles. Motor oil contains zinc, cadmium, and nickel, all of which have been shown to bioaccumulate in low food-chain organisms such as earthworms. Animals consuming the contaminated earthworm also ingest these heavy metals and may be affected.

Roadways and access pathways can also be a conduit for pollutants to enter the environment. Deicing fluids and road salts used in winter leach into waterways, for example. These additives to make travel safer have been shown to decrease survivorship of wood frogs and spotted salamanders in adjacent ponds. Deicing fluid has also been shown to cause skeletal abnormalities in frogs and negatively alter their locomotor performance. This can decrease agility in catching prey and running from predators which increases mortality and survival of the population. Animals are also attracted to the salt added to roads for deicing, increasing roadside mortality. Runoff from sodium chloride ice melt causes eutrophication of water bodies, increasing algal blooms in ponds and streams that receive runoff.

Animals that rely on light patterns and control for biological activities are also impacted by roadways. For example, robins use morning sunlight as a cue to sing and may become confused by highway lights, singing in the middle of the night. Bats are also affected by artificial light and will alter their feeding pathways and flight routes to avoid it. Hatching sea turtles navigate toward the ocean by following sunlight and may mistake car headlights for the route to the sea. When these animals navigate toward roads rather than water, they become stranded and die, impacting the overall population.

Construction projects also facilitate the spread of invasive species. Seeds or burrs from a species may hitchhike on construction equipment and be spread into pristine areas. Mud on vehicles may contain spores or pods which become enmeshed in local soil and cause nonnative plants to grow. These nonnative plants can crowd out local vegetation and alter the local community, causing a ripple effect on vertebrate and invertebrate populations which rely on plants for nutrition. Decreased competition from plants in a cleared area allows invasive species to thrive. Once an invasive species has gaining a foothold, it can be very difficult to remove. For example, buffel grass, originally planted as livestock forage, is now crowding out native species in the Sonoran Desert. This grass can only be eliminated by pulling it out by the roots and requires many volunteers many hours to contain in order to preserve native plant life. It burns hotter than the local grass, red brome, which means that native plants are eliminated in Australia; the cane toad has used roadways to increase its range, crowding out local frogs. When a grasslands fire occurs, the buffel grass survives. The hardy plant does not provide cover or seed for local species and thereby has a negative effect on the local ecosystem. Invasive species are not only plants, but they can also be mammals.

The phenomenon known as edge effect is a negative impact of land construction. When a road is built through a forest, it creates an edge. This edge invites weedy plants and opportunistic roadside species such as mice and voles that would not normally be present in the forest area. This zone of influence may extend into the forest interior, altering the native plant community and local habitats. Shade intolerant plants, as well as weedy plant species, alter the native vegetation and forest composition. This edge habitat is its own microclimate which then impacts the species and plant life present in the forest. Some animals, such as the cowbird, take advantage of the edge habitat and forest fragmentation to further than own range. The cowbird is a brood parasite, laying its eggs in the nests of other birds. The other birds raise the young as their own, crowding out their offspring from the nest and decreasing their reproductive success. Forest birds may experience serious declines in areas where cowbirds have become common as they are not well adapted to their breeding tactics. In addition, opportunistic nest predators such as raccoons and opossums are common in roadside environments. These predators attracted to the area to feed on carrion will also prey on edge nests, decreasing local bird populations.

Another area of potential impact is that of cultural artifacts. In areas with either current or prior indigenous populations, it is important to be aware of any significant land areas during construction. In areas undergoing topographic reformation or soil erosion, care must be taken to avoid impacts that affect land masses or erode culturally significant areas such as trails or footprints. Increases in site access due to construction may also affect culturally significant resources. Access to areas or artifacts must be reduced or avoided if possible to ensure that no artifacts or valuables go missing. Any artifacts found on site must be carefully documented and recorded, along with the appropriate authorities and groups notified. Failure to properly care for a culturally significant site can carry severe penalties and impede construction projects, sometimes permanently. Care must be taken to be sensitive to the needs of the indigenous people for whom the site has significance as well as to be mindful of the greater community, which is also a stakeholder in the area. A sacred landscape or trail can also be negatively affected by construction as it may remove previous visual elements or increase sight lines in a way that was not there previously. This can result in erosion and increased site access which may also negatively affect the site.

Soil and geologic resources in a land area may also be affected by construction. Excavation of topsoil can result in soil erosion, and the removal of rock for road building can disturb the geologic profile of an area. Sand and gravel used for road building generate dust and rock tailings that will impact the local geology, as well as changing the plant profile of the landscape in areas where roads are built. Laying down rock impedes the growth of low-level foliage and reduces the available green space for seeds to spread. This results in forest edges, which reduce the tree cover of any area and available nesting space. Imported rock and gravel do reduce soil erosion in graded areas, however, which keeps current soil intact. If local rock and sand are used, then care must be taken to avoid increasing erosion in areas mined for these resources. Cutting of hillsides for site access can increase instability of slopes and alter drainage patterns. A newly toposcaped area must be carefully monitored after a heavy rain for landslides and earth movement. In areas with soils having special properties, such as cryptobiotic soils, care must be taken to protect the local microbiotic community. The soil microbes are very important to the internal health of the local ecologic community, and altering the soil may cause later impacts. Therefore it is important to keep local soils in place during construction projects and avoid bringing in other soils if possible. Removal of soils will result in nutrient loss and can also affect local water bodies, if erosion in an area is severe. Lost nutrients can be amended (for instance, if the area is to be farmed), but it is generally less costly, environmentally and financially, to keep original nutrient profiles and communities in place.

Visual resources of an area are often affected by construction on land. In sacred areas, this must be avoided if possible. Areas that affect visual resources include access roads, right of way variations, and parking areas. All of these increase open space in an area and disturb the view with light and vehicle pollution. Dust and dirt from local traffic may cause a haze in an area, if it is severe and will need to be managed. The most common way to reduce dust in a construction zone is by water misting from a tanker. Water can be imported or local, and it is important not to spray too much as that will increase erosion by turning the area into mud. Large excavations, such as mining or trenching for electrical lines, can leave visual scars on the landscape which may persist for years to come (particularly in the case of mining). These areas must have reclamation plans to allow the land to recover. In some highway areas where the hillsides have been cut, the local concrete is painted to resemble the rock strata underneath, thereby reducing the visual impact.

A number of mitigation strategies to reduce environmental impacts have been implemented, with varied success. In some areas where roads cross routes of wildlife migration, these roadways are closed during breeding and hatching season. In Florida, this has been particularly successful in increasing the breeding success of frogs and turtles, both of whom are disproportionately affected by roadway mortality (Field et al. 2016). In other areas where snakes are killed while crossing roadways to find nesting sites, construction of nesting boxes has reduced mortality. In other areas with large wildlife, such as deer and bears, construction of wildlife bridges and culverts have been successfully used to reduce mortality. These structures pass over or under roadways, and give wildlife a safe pathway to cross to follow prey and to mate. Bridges are used for large wildlife, while culverts give small animals a way to cross the road safely. In each type of crossing, vegetation is planted on the structure to mimic natural habitat and encourage usage. Culverts are lined with sand or gravel to encourage reptiles to use them. As some animals avoid roadways due to traffic noise, wildlife crossing in areas of lesser vehicle density may see greater usage than those in higher traffic areas. Over time, it has been shown that deer and boars will use wildlife crossing, but that moving through the area does take some habituation. Overall, wildlife crossings on roadways have been a success in reducing animal highway mortality (Noss 2002).

2.1.2 PROJECTS IN WETLANDS

Construction projects in wetlands must include special care to avoid environmental impacts. Wetlands are important to the local ecosystem, as they provide breeding grounds for large and small creatures, conduits for clean water to streams and rivers, nesting spaces for avians, and hiding spaces for prey animals. Wetlands are protected under local, state, and some federal water quality standards, so it is important to be aware of and in compliance with those rules.

Construction can change areas through changes in water quality, water quantity, hydrology, stream channel morphology, and even changes in ground water levels if water is diverted and prevented from reaching the local water table. In areas where mining is part of the construction process, it is important to be aware of local water flows and quality and to have a mitigation plan in place to restore affected water quality at the close of the project. A good example of this is coal mining. Following coal mining, water filling a mine will leach minerals from the surrounding rock, like pyrite ore. Pyrite oxidizes to iron oxide when exposed to air and water, causing the water to acidify and producing an orange sludge. This sludge steals oxygen from the water and can kill the stream if left untreated. Wetlands are actually used as treatment for this acid mine drainage, as cattails trap the iron sludge and their roots help to reoxygenate the water. In areas where the impact is particularly severe, gravel or other limestone containing rock can be added to the water to increase the pH back to values that will support aquatic life.

Construction can also cause impacts by restricting water flows upstream. In areas where the road bed is raised above the land, the roadway will act as small dam. During storm events, this can reroute and restrict water flows, which can have significant impacts downstream. For instance, in the Big Cypress Everglades in South Florida, this restricted flow was found to impact the local wetland, causing water levels to drop and even disappear in some places (Field et al. 2016).

Drainage ditches and culverts can also impact the wetland, causing drainage in areas and altering local habitat. When roads cross a stream, the stream is often diverted or channelized, which can in turn alter received water bodies like wetlands. Culverts and bridges restrict passage of fish and other aquatic creatures that are important to the wetland ecosystems. These creatures oxygenate the water and provide ecosystem balance. Without larger predators, wetlands can become overrun with small prey fish and insects, which affects the local water quality. Channelization of the streambed is also harmful. Creating channels removes the naturally diverse substrate from the streambed which will affect the ecological balance of the wetland ecosystem. This channeling particularly affects the bottom dwelling creatures in the wetland, which depends on nutrients in the streambed to survive. Altering the streambed profile can cause these benthic organisms to die off, therefore causing alterations further up the food chain. If the bottom dwelling creatures survive, they may have difficulty holding on due to the increased current from channeling and be swept downstream of their natural habitat. This type of habitat movement is common with construction in wetland areas and can alter the profile of the natural ecosystem.

Draining of wetlands can exacerbate downstream flooding and cause bank instability. Wetlands provide natural buffers to rising waters and help to absorb pollutants from stormwater runoff. Eliminating these natural stream breaks has been shown to increasing downstream flooding and reduce the stability of streambanks, thereby increasing erosion of the stream channel (Threats to Estuaries 2016). Impacts on wetlands from construction are an important concern as they can have ripple effects throughout the local hydrology. For instance, increased sediment from construction and stream erosion has been shown to have impacts on fishes, decreasing populations (Threats to Estuaries 2016). The extra sediment causes clouding in the receiving waters, making it difficult for fish to locate insect prey floating on top of the water, and reduces oxygen in the stream. The reduced oxygen causes algae and duckweed proliferation which again makes it difficult for fish to locate prey and decreases the water quality. Extra sediment also impacts breeding as fish like salmon use areas of small gravel and rubble to lay their eggs. Sediment can cement graveled streambed in place, impeding oxygen flows and reducing breeding success. During construction, the vast majority of sediment is produced;

however, in the case of an unpaved roadway, sediment will continue to be produced for as long as the roadway remains unvegetated. The quantity of sediment produced during construction of a road, for example, can be as much as 3,000 tons of sediment per mile, as measured during construction of a divided highway in the Scott Run Basin in Virginia (US EPA 1994). In the same study, it was found that construction of the road contributed up to 85% of the sediment within the basin. This was as much as 2000 times that of forest land!

Removal of vegetation during construction can also impact wetland areas. Wetland ecosystems host unique plant life such as rushes and cattails which are natural water filters and promote water recharge. Removal of these plants reduces groundwater recharge and lowers the local water table. Wetland plants increase soil moisture, and in turn support semi aquatic microorganisms and biota which act as support for higher food-chain organisms. Soil strata with reduced moisture due to the removal of wetland plants is prone to greater erosion and reduced community, which in turn decreases the fertility of the soil. Wetland plants provide an important avenue for both nutrient reduction and improvement, such as ammonification and nitrogen fixation which are important for support of crops and plant production. Reduced soil moisture has been shown to increase wind erosion of soils, which can make it difficult for recolonization of indigenous plants should vegetation be allowed to return to the area (Sholarin and Awange 2016).

Wetland ecosystems provide unique transition areas within the local ecosystem and therefore care must be taken to preserve these important plant life and wildlife habitats during construction. Wetlands provide sinks for metals such as iron which can be rereleased into the environment if the wetland area is eliminated or reduced. These heavy metals can have widespread impacts throughout the local ecosystem including increased plant, wildlife, and insect mortality due to exposure. Wetlands are also important nurseries for plant and animals. Fry fish, such as salmon, spawn in quiescent wetlands close to upstream areas upstream, so impacts in these areas can reduce fish populations. Other keystone insects which are essential to healthy streams such as caddisflies, mayflies, and dragonflies make wetlands their home base and provide fodder for aquatic birds. Wetlands provide habitats and important migration grounds for traveling birds like osprey, geese, and herons, which are also key parts of a healthy ecosystem. These communities are important areas of biodiversity which support the overall ecosystem and contribute to a healthy environment. Therefore, it is essential to manage and support construction projects to minimize impacts on the wetland area to protect these important keystone communities.

2.1.3 PROJECTS IN LACUSTRINE/MARINE/ESTUARY ENVIRONMENTS

Construction projects in lacustrine, marine, and estuary environments must be tightly managed to avoid environmental impacts. These ecosystems are delicately balanced and disturbances can upset that balance and cause irreversible damage if caution is not the foremost concern during building. Each of these environments is integral to the soil formation process. Construction can disrupt this process and disturb previous layers, impacting the soil quality over the long term.

Lacustrine environments are areas of still water in lakes that allow layers of fine sediment such as sand, silt, and clay to settle to the bottom and form layers. This process takes many years to unfold and to build up layers of soil on the lake bottom. When the lake is drained, or the area elevated, these thin layers are exposed. The layers in lacustrine soils are very well sorted, and reflect yearly deposition of sediments. Layers are very thin, so disturbance in an area, such as a construction project, can impact many layers of sediment. Lacustrine sediments can be important archaeological and geologic records as they are so well sorted. These sediments are also devoid of coarse particles like gravel so artifacts are readily able to be found.

Marine environments are home to many unique creatures which have habitats on both the water and the land. There is a complex system of regulations that govern construction in the marine environment as it covers both land and water. These regulations come from all sections, including local, state, and federal permitting rules.

The marine environment is also a harsh environment for construction materials, so special fabrications must be used to support structures. The most commonly used materials in marine construction are concrete and steel. Salt corrosion is a significant concern for materials used in marine construction. Exposed steel is subject to deterioration from the current, chemical corrosion from chlorides, and embrittlement from breaking down of the sulfides in the steel. Concrete is subject to many of the same influences, including deterioration and pitting from reactions of oxygen and seawater and flaking from carbonate breakdown. Areas of concrete exposed to the sea must be internally reinforced with steel for structural integrity and to promote longevity of the structure.

Estuaries are also areas where environmental impacts during construction must be carefully monitored and minimized. An estuary is a transitional environment between freshwater and saltwater, and therefore a very important keystone in the ecosystem. Due to the transitional nature of these areas, many unique species call them home. If the estuary is disturbed, these species become displaced, either relocating or in extreme cases dying out. For many years, estuaries were not treated as important to the health of the environment. Now, however, it has been recognized that these areas are integral to species survival and they are protected. Construction projects can negatively impact estuaries by draining or dredging the area, or filling it with sediment. These impacts must be avoided in order for the estuary to remain healthy.

Sediment is a particular concern for estuary health. Clearing land near an estuary increases the sediment load, which can have far reaching impacts within the estuary. Cleared land erodes more quickly than vegetated land, so when a storm occurs, the estuary must cope with an increased sediment load. Increases in sediment in the estuary also increase the water turbidity. Turbidity is the water clarity. In estuaries, the water is typically quite clear, which allows aquatic organisms to thrive and for high rates of oxygenation in the water. Increased turbidity from stormwater that carries a high sediment load can submerge bottom dwelling creatures and make it difficult for others to locate food and breed in the murky water. Changes in the nature of sediments can also affect the health of the estuary. Some creatures such as caddisflies and mayflies prefer small particle sizes of sediment to large ones. They use these small sediment particles to make their homes. Increases in muddy sediment with large particles that enters the estuary following land clearing will decrease the number of caddisflies and mayflies as there is more competition for sediment particles. This decrease in insect species can have ripple effects along the food chain by increasing competition among fish and avian species.

The rate of sedimentation in an estuary is also important. Due to their transitional nature, estuaries often experience varied rates of sedimentation, depending on their location and the time of year. For instance, those areas closer to the shore have greater rates of sedimentation, while those closer to the sea tend to be areas of deeper water. During storm season, the estuary acts as a buffer for the area, helping to mitigate the impacts of storm surges. This stormwater stirs the sediment in the estuary and can help clear areas of stagnation so that the sediment can be replaced. When the area around an estuary is cleared, runoff from soil erosion can increase sediment loads and the rate of infill in the estuary. This rate of infill can exceed the natural capacity of the estuary, effectively burying it and removing this area of productive ecosystem. Infill and sedimentation in the estuary can also impact the plant and animal communities. Some plants and animals prefer deeper water, which can disappear when sediment accumulates within an estuary. These plants and animals will then have to migrate further out in order to survive and may not be able to cope with currents or increased salt loads in the remaining areas of deeper water. Changes in plant community can make it difficult for those unique animals that rely on the diverse estuary flora to continue to populate the area. It is also important to note that impacts on the estuary flora can change the overall community of the estuary, which can upset the ecosystem balance. Losses of sea grass beds, the spread of mangroves, and the decrease in shellfish beds are all losses of habitat that have been linked to infilling of estuaries from sediment runoff due to construction projects (Saenger et al. 2013).

2.2 LONG-TERM IMPACTS

2.2.1 WILDLIFE IMPACTS

Long-term impacts on wildlife from human activities are far reaching and significant. In fact, human activity in the single largest leading threat to biodiversity. Impacts on wildlife take three main forms: fragmentation, degradation, and outright loss of habitat. The impetus for these actions is that of human ecology; the need or want for the human population to expand into undeveloped areas. Only if effective conservation strategies are adopted can significant impacts be avoided.

Habitat fragmentation is a significant cause of wildlife impacts and habitat loss. Fragmentation of habitats occurs when a large ecosystem is broken up into smaller parcels which are isolated from one another by human activities. This is an umbrella term that describes the complete process of habitat loss rather than simply the remaining patchwork of habitat. As such, it is important to understand that fragmentation is part of a larger process of ecosystem degradation and not merely a symptom of human activity. Changes in ecology and habitat have many factors which shape the health of the overall environment. Human impacts are an important factor, but outside influences such as climate change can also drive fragmentation of an area. It is important to note that habitat fragmentation is a landscape-level phenomenon, and processes in fragmented zones (area, edge effects, and shape complexity) can only be understood within a landscape context (isolation and matrix structure).

A dominant factor of habitat loss is increased loss in fragmented areas. For example, an area that is first split in half is more prone to being split further into smaller areas, much like a cookie crumbling once broken in half. Each area is less strong the smaller it becomes, and the loss accelerates. Decreases in fragmented areas result in declines in population density and species richness, and significant alterations to community composition, species interactions, and ecosystem functioning. Communities that become fragmented are often skewed with regard to the predator–prey balance, as there is increased competition for resources. As such, prey populations may increase if there are insufficient predators to keep them in check, or predators may deplete an area beyond its ability to recover naturally. This affects the overall ecosystem function, forcing species migration and creating further fragmentation of the area.

Fragmented areas become microcosms of the greater ecosystem, increasing competition for space and resources within the area. The boundary of a fragmented zone can be fluid depending on its position within the ecosystem and surrounding area. Avians and large predators can cross many fragmented areas while roaming, allowing for some connection between fragmented areas. As such, there can be some cross over between predators and prey once zones become disconnected, but the natural balance is skewed toward those that survive the journey between sectors. This leads to decreases in genetic diversity and the overall population declines. Boundaries in fragmented areas are not only defined by vegetation but are also influenced by the distances to other fragmented areas and areas of human influence.

The impacts of boundaries have two distinct components: the distance between the edge of the fragmented area and the overall landscape, and the magnitude of the influence of that distance when contrasted between the fragment and the overall landscape. The greater the distance between the edge of the fragmented area and the overall landscape, the more likely it is that the fragmented zone will experience population declines. Isolation of fragmented areas reduces connectivity of populations and therefore contributes to decreases in genetic diversity. Populations with low degrees of genetic diversity have difficulty adapting to changing conditions such as changes in vegetation and hydrology. For example, outbreaks of the Hendra virus in flying foxes in Australia have been shown to be more lethal in urban populations which are isolated than in rural populations where low-level outbreaks lend immunity to the herd (Plowright 2011). This is not isolated to mammalian populations either. In a study of the herb ribwort plantain, researchers found that powdery mildew, a leaf blight, disproportionately affected those plants in isolated populations (Plowright 2011). Plants in isolated populations were more likely to become infected, and those in larger, connected patches were more likely to resist infection. Researchers noted that regular exposures seemed to heighten plant defenses, serving almost as evolutionary inoculations against the leaf blight. In large patches, disease-resistant plants were more likely to reproduce and spread, and—crucially—there was a larger gene pool from which disease-resistant mutations might arise. Small-patch plants, less-exposed and with a smaller gene pool to draw from, remained susceptible to infection when it finally did occur.

The surrounding landscape composition has a dominant influence on population dynamics, species diversity, and ecosystem processes in fragmented areas. If the ecosystem is largely healthy and fragmentation is small, those fragmented areas are less likely to suffer from large-scale effects on population and species survival. Species diversity is more easily maintained, as is community diversity. Ecosystem processes are not as easily disrupted, which also contributes to greater survival and population balance. It is also important to note that changes in the overall landscape may not have initial effects on the fragmented areas, but effects may become apparent over longer time scales. For instance, construction of a road through an area may cause fragmentation, but implementation of a wildlife bridge may mitigate the effects and help separated populations to recover. Conversely, a species may appear initially unaffected by fragmentation, but over time reduced breeding success and changes in vegetation may severely reduce its numbers causing extinction.

There are many factors which influence the effects of habitat fragmentation, both internally in patches and externally in the overall ecosystem. Examples of these factors include components of global environmental change, species invasions, land-use intensification, and climate change. As each of these factors interlock, it is important to quantify effects both on an overall and individual scale. Qualitative effects such as esthetics and edge habitats zones also need to be taken into account. Only when all areas are considered holistically can the impacts of habitat fragmentation be accurately assessed.

Degradation is another important avenue of human impact on the environment that has long-term effects. Degradation refers to the decrease in habitat quality for wildlife due to human activities. Degradation is caused by construction such as road building or land clearing for human use. Degradation can also occur in fragmented areas where humans encroach on wildlife habitat. Mixed use areas, such as highways or parklands, accelerate habitat degradations by increasing human activity around wildlife habitats.

Fragmented areas are subjected to greater and more accelerated rates of degradation by human activity. This is due to the already reduced space within the fragmented patch of habitat. The more complex the fragment is, the greater the reduction in core habitat area available for habitat. A more complex fragment is affected more by edge effects. Edge effects are changes in plant community (such as grasses at the edge of a forest road), increased animal mortality (such as an increase in individual roadkills from animals attempting to reach another fragment), and changes in community composition (such as gender or predator–prey disparities due to decreases in breeding success). Availability of core habitat area reduces these negative effects and allows the ecosystem balance to continue as normal.

Parkland areas can increase the rates of survival for wildlife species, but do have drawbacks associated with their management and impacts. One area that is important for degradation is that these areas have differing regulations from those outside the parkland boundaries. These surrounding land use categories such as grazing and farming may degrade the area by outside processes such as using water resources and the invasion of nonnative seeds in the parkland area. Migratory corridors are among the areas that are most adversely affected by degradation. These areas are easily degraded by human activity as they are not typically protected and can cross multiple zones of human influence. Human influences on these corridors can thereby affect breeding success and species survival.

Habitat degradation can also come from within the protected zones. In parkland areas, there are often trails and other zones of human use. While not as large as a road or a building, these encroachments nonetheless have degradation effects. Trail construction results in small-scale edge effects like plant population changes within the forest floor. Small mammals such as mice and voles will alter their routes for gathering seeds around trails which can cause a ripple effect through the forest. Predators hunting these small prey will follow them, resulting in self isolation and fragmentation of species. Trails effectively concentrate animals within a forest area without the large-scale construction that is typically associated with habitat fragmentation. The areas around the trail become degraded as human feet and hands remove vegetation from the pathway and build up rocky areas to provide better pathways. Careless hikers can leave debris and food waste in their wake that accelerates the degradation of an area.

In areas of high population growth, pressure on the borders of parkland area can also lead to degradation. Increased pressure on the border results in predation of edge resources such as wood and water to support the population bordering the area. This "boundary nibbling" erodes the community within the protected zones by degrading their habitat through inadvertent use of resources.

Agriculture is a large driver of degradation by humans in protected areas. Concentration of livestock in small areas leads to overgrazing, silting of water bodies, and soil erosion, which degrades the surrounding wildlife habitat. Overgrazed areas take longer to recover than managed areas, again decreasing the usefulness of the surrounding area to those natural communities. Soil erosion reduces healthy soil microbial communities, making it difficult for native plants to grow and retain footholds in an area. Lack of healthy plants degrades habitat by reducing areas for prey to hide and herbivores to graze. Water bodies proximal to protected areas and grazing land are also disproportionately affected by degradation. Water scarcity in these areas leads to drying up of effluvial ponds and streams, and decreases in water quality can drive degradation. Natural processes in water to replace nutrients and maintain usability are often slow, which means that, like plants, water bodies take a long time to recover from degradation and the effects of human activities.

Poverty and demographic effects are also large contributors to habitat degradation. Poverty drives individuals to seek resources at the cheapest labor and material costs, such as encroachment into protected areas where resources are easy to obtain, even if illegally. In the Serengeti, population expansion has led to the decrease in availability of land for grazing livestock, which has in turn lead to habitat degradation as agriculture competes for decreasing resources within the area. Human expansion has also lead to degradation of edge lands along protected areas, and in some regions, such as Maswa, to the redrawing of borders to allow human encroachment into protected lands. Minimal enforcement of area borders also contributes to habitat degradation as authorities turn a blind eye to grazing and gathering of firewood in these areas. These illegal activities deplete the area of resources and degrade wildlife habitat.

Large-scale projects such as irrigation and hydropower degrade an area by decreasing the availability of resources. Irrigation projects which reduce the need to build roads to bring water to agriculture simultaneously fragment and subsequently degrade areas by cutting channels through protected lands to access water resources. Similarly, hydropower projects reroute spawning pathways for fish by damming rivers and streams. Construction of lakes behind dams encourages human use and expands areas of degradation along water bodies. Use of resources such as hydropower and irrigation also reduces available resources within the protected areas to which they may be connected, again resulting in degradation. Another example of degradation is the use of fertilizer for crops, which can runoff in rainstorms and degrades nearby water bodies within protected areas. Eutrophication of streams and ponds affects wildlife communities by degrading the quality of the local water resources.

Wildlife are also negatively affected by outright loss of habitat. In this instance, habitat becomes so degraded that it is no longer usable by its inhabitants. In other instances, habitat that has been previously fragmented is absorbed into industrial or agricultural usage, again rendering it unsuitable for the residence of wildlife. The consequences of outright habitat loss are cumulative over the long term, causing instability and deterioration of the greater ecosystem. Even loss of tree cover while maintaining ground cover is considered outright habitat loss as this removes a key portion of the avian community from habitation of the area. Loss of even a single species can have a ripple effect throughout the ecosystem as other species which prey on those animals diminish due to lack of secure food sources.

Although previously habitat losses were not the most important factor in decline of species, being overshadowed for centuries by overexploitation and introduction of exotic species, the relative importance of outright loss has increased in recent decades. In fact, habitat loss has emerged as the most severe threat to biodiversity worldwide threatening some 85% of all species classified as "threatened" and "endangered" by the World Conservation Union (Kideghesho et al. 2006). It is the most pervasive to birds, mammals, and amphibians. It affects 86% of all threatened birds, 86% of threatened mammals, and 88% of threatened amphibians, according to recent statistics.

Outright destruction or degradation of habitat beyond the ability of wildlife use can have severe long-term impacts. These impacts include species extinction, ecosystem collapse, and degradation of the area beyond its ability to support life, such as desertification and eutrophication. It is very difficult for an area to recover from these effects without substantial remediation and reintroduction practices. One of the key difficulties of outright habitat destruction is the loss of nutrients, both in soil and in water. Amendments can be applied to replenish the area, but without flora and fauna, the process is not sustainable. A good simple example of this is the use of crop rotation by farmers to restore depleted lands. Rotation of crops allows for different nutrients to be amended and utilized by plants, which gives the ecosystem a chance to recover from overexploitation. Even so, nitrogen- and phosphorous-containing fertilizers are often needed to amend the available soils as the revitalization process is not quick enough to restore a field within a growing season. For farmers, this becomes a costly and continuous process of fertilizing as the area is never given time to fully recover. In order to achieve results, farmers often must over fertilize land, which leads to the deterioration of nearby water bodies from storm water washout of excess fertilizer material. In order for nutrients to be replenished, land must lie fallow to fully recover. In areas that use farming as their primary means of support for humans, this is often not possible, so the land becomes more and more depleted over time until it is eventually uninhabitable. This outright loss of habitat leads to migration of wildlife and degradation of the overall ecosystem habitat.

Habitat loss also affects migratory corridors. Animals large and small typically migrate in response to ecosystem pressures such as isolation and food stressors. Loss of these migratory pathways further diminishes the ability of the migratory species to survive and to reproduce successfully. Even if the available area around a primary habitat is protected, loss of migratory corridor habitat can cause a species to become endangered due to lack of breeding success or genetic variety. Depletion of available nutrients is also a factor in long-term wildlife survival; isolated species have been shown to degrade over time as their primary areas become overgrazed or over preyed upon. Some scientific studies have led to predictions and generalizations on the ecological impacts of isolation and small ecological units. Predictions for loss of large mammals in East Africa suggest that when the areas are isolated for 50, 500, and 5,000 years without intervention of scientific management, the smallest reserves may lose 23%, 65%, and 88% of the species, respectively (Kideghesho et al. 2006). In contrast, the risk is lower for the largest reserves. They may lose 6%, 35%, and 73% in the respective time intervals. Likewise, extrapolations from estimates for habitat loss have led to the most widely quoted generalizations that the loss of 90% of a local habitat results in loss of half of the species present! The extent of this loss cannot be overstated as cornerstone species such as avians and amphibians are disproportionately affected by losses and those losses have ripple effects throughout the ecosystem.

Human population pressure is cited as one of the main contributors to outright habitat loss, mainly through deforestation prompted by increased demand for arable land, settlements, and fuelwood. The majority of the Sub-Saharan Africa population is dependent on fuelwood: 82% of all Nigerians, 70% – Kenyans, 80% – Malagasies, 74% – Ghanaians, 93% – Ethiopians, 90% – Somalians, and 81% – Sudanese, making this problem particularly acute on the African continent (Kideghesho et al. 2006).

Similar to habitat degradation, poverty is a significant factor driving habitat loss, which exacerbates the effects of overgrazing, deforestation, and settlement. Areas with high degrees of poverty are typically those that are also least exposed to sustainable practices for agriculture and settlement, as these populations are focused on the efforts of daily survival rather than the long-term health of the area. These pressures are exacerbated by high population growth which requires greater inputs of resources. As these low-income areas are least able to access technology and funding to maximize production, expansion into new lands becomes the default solution to produce supporting resources. These lands are often sensitive areas, including migratory corridors, dispersal areas, and breeding grounds for wildlife. Even in protected lands, expansion is a problem as borders are not well patrolled or monitored. Without adequate enforcement, these areas are just as vulnerable to degradation and habitat loss. Furthermore, carrying capacity in these areas is already diminished due to the use of antiquated agricultural processes and equipment. Even in more affluent areas with access to electricity, fuelwood is still required by the population as electricity use is prioritized for lighting and communication rather than heating or cooking. Only by providing adequate infrastructure can this stress be alleviated, and that requires large inputs of funding and outside efforts.

2.2.2 WATER/MINERAL RESOURCE IMPACTS

Long-term impacts on water and mineral resources are not possible without proper management of construction projects. The main impacts of projects are water pollution and increased water stressors within the area of the construction project. Mineral resources impacts are linked to water resource changes, as mining and mineral refining use large amounts of water and heavily impact local water quality.

In order to achieve the lowest impact on a site, it is important to implement a robust water quality monitoring program. Monitoring programs conserve water, protect the quality of water resources, and also support and protect the greater ecosystem. By having a robust monitoring system in place, effects of projects can be easily seen, tracked, and effects minimized. The existing water quality baselines can be established in order to provide information for regulatory and environmental compliance metrics. When projects are started within the area, this information is also important as it can help indicate which parameters are worthy of attention and continued observation. Common water quality metrics are pH, conductivity, and turbidity. Other measurements can include total suspended solids and heavy metals, which are more commonly monitored in conjunction with mineral resources projects rather than traditional building sites, but can be important metrics in either case.

The most common water quality impacts resulting from construction projects typically arise from erosion and stormwater runoff. Both these events commonly result in decreased water quality in the local area and can have impacts further downstream, as well as long term. It is important to establish baselines for monitoring during both dry periods and weather events to be able to monitor changes in water quality once projects have begun. Impacts from stormwater may only be evident during periods of heavy rain so it is important to also monitor during these events to generate a comprehensive picture of the local water quality.

Mineral extraction projects require complex and long-term water quality monitoring as effects from mining may not be seen until years after the project has concluded. Mining and mineral usage parameters can directly affect water quality through deterioration of water quality from exposure to waste rock and tailings. This can result in mine drainage impacts and increases of heavy metals in local waters which decrease local water quality and degrade the overall environment.

It is important when establishing monitoring of a site that samples describe the overall area well. This means that samples must be taken both upstream and downstream of the area as well as including any tributaries if the project may potentially affect adjacent regions. For large-scale projects, this is important for regulatory compliance and monitoring as well as rehabilitation of the area once the project is finished. As part of monitoring, the site should also be visually inspected during sampling in order to help pinpoint any problems that may arise.

As most large impacts occur during times of site discharge, it is important to take samples both upstream and downstream of the area as soon as is practical following rainfall events. Samples taken upstream should be outside the impact of the site area, but as close to it as possible to provide the best guidelines for original water quality. Areas selected for sampling should also have permanent water and be flowing, if possible, to provide the best assessment for the area. Ephemeral pools or ponds are not to be used as upstream measurements as this will skew natural results. However, these areas should be included within the realm of sampling sites, particularly if they occur due to site discharge or stormwater runoff from the site. Downstream monitoring samples must be undertaken within influence of the discharge of the project where there is sufficient mixing to show the representative impact of any site discharge on the receiving waters, unless this is not possible due to safety issues or if suitable sampling sites are not available.

For some projects, there may be external discharges which occur between the upstream and downstream sampling locations that may impact the receiving water quality. These may include other tributaries or stormwater discharges from agriculture or industry. In these cases, the simple comparison of upstream and downstream sample results may not indicate actual project impacts. Therefore, sampling locations must be selected to obtain results that are indicative of the project impacts relative to the other sources. Other variables between the sampling locations to be considered include the depth of water, the flow velocity, resuspension of sediments, and stream configuration, all of which may impact the data and should be factored in during interpretation.

Site runoff is typically the largest contributor to erosion of soil from the site as well as the most common avenue of pollution of a local water body from a construction project. For most projects, monitoring on a regular basis with a portable probe is sufficient to meet established guidelines. These measurements should be carefully recorded and stored for reference before, during, and after completion of a project. The probe should be calibrated regularly and users trained in proper measurement techniques. In some cases, where there is continuous discharge to a receiving water body, *in situ* monitoring is warranted. *In situ* monitors provide continuous data regarding water quality that help to show the conditions of a site and the impacts that an adjacent project has on a local water body over time. Continuous monitors help to yield a picture of the changes in water quality that can result from regular project discharges. In mining projects, these types of sampling programs are more common, as discharge is often continuous, where in construction projects, it is most typically isolated events that coincide with stages in project completion or rainfall events. Either way, appropriate monitoring techniques and intervals should provide project management with a comprehensive picture of the site of water quality before, during, and after the project.

Monitoring of water quality may results in the implementation of other engineered controls. Examples of these controls include construction of catchments and swales for stormwater management and revegetation of disturbed areas. Once an area has been shown to stabilize, monitoring may decrease or stop altogether.

For construction in areas with minerals resources, there are many considerations. Most mineral-rich areas are in remote locations, which necessitates the transportation of humans and heavy equipment to the site. Roads must be built, and supplies transported to site, all of which impact the local environment. Impacts on the resources themselves are also possible as trace minerals can be lost during extraction and transport.

Overall, the extraction of mineral resources is one of the most impactful projects on the greater environment. Mining results in pollution, land subsidence, deterioration of the surrounding environment, decreased water quality, and increased energy usage. Mineral resources are then used to support industry and produce energy, which in turn causes climate change.

Pollution is one of the main results of mineral extraction and construction projects for mining. Pollution results from both extraction and processing of minerals and affects both water and air. Local water sources become contaminated by leaching from waste rock, as well as washing processes necessary to produce pristine minerals for industry. These leaching processes can cause heavy metals to contaminate ground water and surface water. Ponds are often constructed on site to contain water leached from mine tailings, but this construction is not typically water tight and seepage into outside sources can result. Stormwater runoff from these sites is also an avenue for contamination of local water sources.

2.2.3 SUSTAINABILITY OF LIFE IMPACTS

Sustainability of life within an area may be impacted by construction projects. Sustainability of life is defined as enhancing the quality of life by allowing people to live in a healthy environment as well as improving social, economic, and environmental conditions for present and future generations. Given the large scale of construction (accounting for 11.8 million job in Europe in 2015), as well as the large amount of money involved (10% of GDP for Europe in 2015), it is natural that sustainable pathways be secured to protect future investments (Ortiz et al. 2009). This section is also responsible for one of the largest zones of environmental impact from industry, so application and assessment of sustainable principles are a good fit for this area.

One of the most prominent strategies to overcome historically poor resource management within the construction industry is life cycle assessment. Life cycle assessment, or LCA, examines the entire environmental load of processes and products used during construction. This is called cradle to grave evaluation and encompasses the processes from raw materials to landfill of materials used in construction. LCA can be useful to determine where the greatest environmental impacts are found and how they can best be reduced.

The LCA contains information associated with the acquisition of raw materials, energy use, content of materials and chemical substances, emissions into the air, land and water, and waste generation. This is done for each material used in construction and can then be imported into the overall environmental assessment and monitoring programs to help improve the sustainability of life for the project. Coupled with eco-design (the relationship between a material and the environment), LCA is a powerful tool to help improve sustainability of life before, during and after a construction project.

Additional strategies for sustainability of life during construction include incorporation of recycled and reused materials as part of the building. Recycled and reused materials reduce environmental and energy loads by reducing the amount of raw materials used in production. For example, using recycled concrete during construction reduced energy loads required by nearly 15% as well as impacting the generation of carbon dioxide and greenhouse gases (Ortiz et al. 2009). Smart material choice during construction, while having greater initial cost, can impact the environmental consequence of the project by reducing later maintenance costs as well as energy costs. A good example of this is the inclusion of nanomaterials, such as titanium dioxide, within building cement. The titanium dioxide nanomaterial is self-cleaning due its strong oxidizing potential, reducing maintenance costs, as well as increasing the longevity of the structure.

Sustainability of life during construction can be improved and impacts decreased by using low impact methods of materials transportation and sourcing. Local timber, insulation, and water can be sourced in order to reduce the transportation footprint of the project, while increasing the sustainable potential of resources within the area by making smart choices. For example, using fast growing bamboo for framing timber rather than hardwood can support the sustainable wood industry and provide a lower carbon footprint example for other new construction in the area.

Material disposal is also an important avenue for sustainability of life during the construction and usage of the project. Almost 15% of building material becomes waste during construction of a typical industry structure (Ortiz et al. 2009). This value encompasses timber, insulation, and concrete; in fact, nearly all aspects of construction have some form of waste material associated with them. Life cycle analysis can help to identify common avenues for waste within projects and reduce that waste by redirecting waste flows and reusing materials. Purchase of new materials for individual projects can also be avoided by evaluating materials flows and directing construction such that bulk materials can be purchased and used to support multiple projects rather than transporting individual sets to construction sites. This project management strategy can also be applied to the transportation of material before, during, and after construction to reduce carbon dioxide impacts by minimizing transport between sites. Material managed in this way can also be sustainably sourced and purchased, which will in turn reduce the environmental impacts of the project. Moreover, waste generated during and after construction can be centralized which will also reduce the impacts and increase the sustainability of life within the project sector. Smart materials sourcing can then include recycled and reused materials to better support sustainability and life cycle impacts within the project.

CHAPTER 3

AGRICULTURAL PRACTICES MANAGEMENT

Agricultural practices management is an area of significant research into best practices and environmental stewardship. Management of agriculture is important for sustainable practices as well as maintenance of the environment. The continual ability to conduct agriculture on local land is directly tied to good management practice (McLaughlin and Mineau 1995). Agriculture is inherently harsh on the environment. Tillage, drainage, grazing, and usage of pesticides and fertilizers can all have significant effects on flora and fauna. In fact, unsustainable agriculture has been tapped as one of the most environmentally damaging human processes to occur within the last 100 years (WWF 2017).

The general trend to more industrialized farming and aquaculture can have a profound impact on soil community, biodiversity, and water quality. Proper management techniques, including engineered controls such as water testing and crop rotation, can help to minimize these effects. Agriculture has also been used in pilot programs for the biofuels industry, adding to the importance of proper management techniques. Management of agriculture involves many stakeholders including economic, political, and technological interests, making management even more complex. Recently, the adoption of best practices techniques has sought to clarify the roles of these stakeholders in agriculture management and establish base guidelines for all to follow.

Best management practices, or BMPs, for agriculture date to 1977, as part of amendments to the Clean Water Act (Rawson and Moltmann 1995). They were originally established to provide soil conservation guidelines that also had water quality benefits. Techniques originally outlined in the 1977 requirements include cover crops, erosion control, and soil testing (Rawson and Moltmann 1995). Over time, these guidelines have been expanded to cover soil and air quality as well as adapted by conservation officers within individual states to cover instances unique to those areas (such as irrigation techniques in Arizona). Management practices are meant to be highly adaptable and as such occur as broad guidelines rather than enumerated techniques. Examples of BMP guidelines include integrated pest management, techniques for pesticide application as well as safe chemical usage.

BMPs are meant to provide benefit to the agricultural operation while minimizing the harmful effects farming has on the local environment. For example, farmers in the Chesapeake region of Maryland have a set of BMPs that specifically address streambank protection from erosion as this is a locally concerning issue (Chesapeake Bay Foundation 2016). Farmers in this region encourage the growth of streamside buffers by planting native trees and grasses along streams that run through pasture area as well as putting up fencing that keeps livestock out of the stream. These BMPs help minimize erosion, runoff from agricultural waste, and reduce water-borne disease among cattle. The Chesapeake Bay Foundation has estimated that adoption of their BMPs could reduce nitrogen pollution into the Chesapeake Bay by almost 60%, which is significant increase in water quality for the region (Chesapeake Bay Foundation 2016).

Another example of management practice adapted to suit the local region is the set of BMPs in the San Joaquin Valley (Kings River Conservation District 2008). These BMPs are a set of stewardship efforts that monitor chemical water quality parameters that address non-point source agricultural pollution combined with physical operation changes that aim to reduce water usage and protect water quality. There are also elements that address soil management, and pesticide and fertilizer application. As the San Joaquin Valley produces more than 250 varieties of crops that are water intensive, these BMPs help promote sustainable production and ensure that agriculture will remain a viable industry for the future. Examples of physical operation changes that reduce water usage are field leveling (in which farmers reduce soil erosion by reducing the velocity of applied water; a level field does not allow water loss via runoff), conversion to high efficiency water application systems such as drip or micro-irrigation, and tail water returns (in which runoff is captured and applied to irrigate the next field in line). The San Joaquin Valley BMPs also provide a list of soil and sediment management practices which farmers have used with great success in reducing erosion and protecting soil quality. These BMPs cover such topics as conservation tillage (leaving previous crop residue on the soil surface to reduce direct soil erosion), vegetated buffers and ditches (which slow water runoff and reduce degradation associated with nutrient runoff and soil loss), and use of cover crops.

Pesticide and nutrient management practices are also important elements of best management practices. As many farmers use chemistry to promote higher crop yields and reduce natural attrition of plants due to pests and nutrient availability, this area of management is an important one. Pesticides to reduce infestations of crops and increase yields within agriculture have been a management technique for thousands of years. The first recorded use of insecticides was 4500 years ago, by Sumerians who used sulfur compounds to control insects and mites (IUPAC 2010). Pyrethrum, which is derived from the dried flowers of *Chrysanthemum cinerariaefolium* "Pyrethrum daisies," has been used as an insecticide for over 2000 years. Persians used the powder to protect stored grain and, later, Crusaders brought information back to Europe that dried round daisies controlled head lice. Many inorganic chemicals have been used since ancient times as pesticides, including *Bordeaux Mixture*, which is based on copper sulfate and lime, that is still used against various fungal diseases today (IUPAC 2010). Until 1940, many pesticides used in agriculture were organic in nature, being derived from natural sources or sourced from by-products of coal and industrial processes. Early organics such as nitrophenols, chlorophenols, creosote, naphthalene, and petroleum oils were used for fungal and insect pests, while ammonium sulfate and sodium arsenate were used as herbicides, each being derived from either coal production or related industrial process. The advantage of these products was that they were widely available and could be used cheaply. The drawback for many of these products was their high rates of application, lack of selectivity and phytotoxicity. Early management practices around these pesticides address their rates of application and suggest ways to reduce usage (IUPAC 2010). The growth in synthetic pesticides accelerated in the 1940s with the discovery of the effects of DDT, BHC, aldrin, dieldrin, endrin, chlordane, parathion, captan, and 2,4 D. These products were effective and inexpensive with DDT being the most popular, because of its broad-spectrum activity. DDT was widely used, appeared to have low toxicity to mammals, and reduced insect-borne diseases, like malaria, yellow fever, and typhus. In 1949, Dr. Paul Muller won the Nobel Prize in Medicine for discovering its insecticidal properties and it was widely used in agriculture almost from the moment of discovery. In this era of early pesticide use, those in agriculture were mainly unconcerned with management of pesticides as many were cheap, highly effective against pests, and the local environment had yet to show adverse effects from the use of these chemicals. In 1946, a study was published showing the resistance of house flies to DDT, but not much attention was paid to the subject of pesticide management until indiscriminate effects of pesticide usage were highlighted by Rachel Carson's 1962 book, *Silent*

Spring (IUPAC 2010). The development of pesticides and insecticides continued in the 1970s and 1980s with the advent of the world's best-selling herbicide, glyphosate, as well as the pyrethroid insecticides and imidazole fungicides. These chemical agents have a single mode of action, however, and resistance soon developed.

Modern techniques of increasing crop yields by reducing weeds and insects all fall under the umbrella of integrated pest management or IPM. IPM uses a combination of insecticides and herbicides which discourage pest development as well as intelligent application techniques that use small amounts of agrochemicals to eliminate problems before they start. By spraying at designated times in crop and pest life cycles, the amount of chemicals can be minimized with maximum effect. Integrated pest management also takes into account beneficial insects such as ladybugs and encourages their breeding success. This strategy uses good insects to suppress the damaging populations and provides for an overall healthier crop. Biological controls like encouraging natural enemies of pests have been shown to be effective ways to control pests when combined with chemical controls like pesticides (University of California 1996).

Application of pesticides is another aspect of IPM. New technology has been able to reduce the amount of pesticides sprayed in a field or orchards by taking advantage of innovative detection techniques that sense a tree or weed and only spray pesticide when vegetation is present (Kings River Conservation District 2008). Other technology uses electrostatically charged pesticides that are attracted to foliage for pesticide application.

Cultural and mechanical controls are additional parts of integrated pest management that can be used as effective methods of managing pests. Cultural controls are practices that discourage pest establishment by disrupting survival. For examples, changes in irrigation can drown pests or cause them to die of thirst. Mechanical controls are controls that are generally used for larger pests such as mice or voles. These types of controls include fencing, traps, and barriers.

Fertilizer and nutrient management is another important aspect of best practices for agriculture. All best practices guidelines have some component of nutrient management as this was in the spirit of the original 1977 Clean Water Act establishment of BMPs. Nonpoint source pollution reduction of ground water and surface waters is an important management area that is continually evolving. Effective nutrient management also allows growers to apply less fertilizers while maintaining current crop yields. Aspects of nutrient management include management of irrigation, establishment of buffers, conservation tillage, and erosion and sediment control.

Irrigation management is one of the most important areas of best practice for nutrient management. Irrigation systems that minimize water usage and retain overflow play an important role in reducing nonpoint source pollution in surrounding waterways. Micro-irrigation and drip systems are increasingly used in the water-scarce areas of the West such as California. These systems flow water at a slow rate to the roots of the plant rather than overhead spraying of the leaves, which is common in the Midwest. This technique reduces evapotranspiration of water and helps to retain fertilizers as there is minimal runoff and therefore minimal water losses. Retention of overflow water from irrigation is another important nutrient management technique. It has been estimated that over 20% of nitrogen applied to fields is lost through irrigation water runoff (Unlu et al. 1999). Overflow water retention, also called tail water return, captures runoff from irrigation fields and returns it for irrigation of the next area. Farmers using this best management practice reduce their water and nutrient usage. Capturing runoff in settling ponds is an additional aspect of irrigation management that reduces nutrient loads. In these ponds, flocculants can be added which coalesce soil particles so they are retained at the bottom of the pond. This practice has been shown to reduce suspended solids, nutrient, and fertilizer loads by 50 to 90% in receiving waters (Kings River Conservation District 2008).

Irrigation scheduling is another important best practice area for nutrient management. Information systems that use weather stations to monitor precipitation and evapotranspiration provide farmers with data that allow them to determine best times during the year to irrigation. This data can be aggregated with crop growth data to give the best information regarding watering. In addition, water stress is a great management tool for certain crops to spur plant growth in certain directions (more reproductive growth, less vegetative), which can also reduce the need for additional fertilizers and nutrients.

Monitoring and sampling of crops and soil are also an important part of fertilizer and nutrient management. Nitrates and phosphates are commonly sampled as well as soil pH and occasionally microbial communities. Effective monitoring of soil and water can better inform farmers about their soil and resultant crop health and help minimize the amount of fertilizer and nutrients applied to a crop. Of applied nutrients, nitrogen is the most concerning as it is water soluble in nitrate form and therefore most easily lost. Leaching of nitrate from agricultural fields can elevate concentrations in underlying groundwater to levels unacceptable for drinking water quality. In the Suffolk County area of Long Island, for example, almost 10% of private wells tested for nitrate exceeded the 10 mg/L drinking water standard (Trautmann et al. 2012).

Effective nutrient management is also a matter of accounting for all nutrients within the system, which is done most commonly through monitoring. Proper nutrient management economizes the natural process of nutrient cycling to optimize crop growth and minimize environmental impacts (US EPA 2016a). There are 16 elements that have been identified as essential to plant growth: carbon (C), hydrogen (H), nitrogen (N), oxygen (O), phosphorous (P), potassium (K), sulfur (S), calcium (Ca), magnesium (Mg), iron (Fe), copper (Cu), zinc (Zn), manganese (Mn), molybdenum (Mo), chlorine (Cl), and boron (B). Carbon, hydrogen, and oxygen are products of photosynthesis, and the other elements are most commonly present in soils. Nitrogen, sulfur, phosphorous, calcium, magnesium, and potassium are needed by plants in larger quantities and are therefore classified as macronutrients and the others are needed in smaller quantities and classified as micronutrients. Macronutrients are also added to soils as fertilizer and amendments to encourage production, while micronutrients are generally sufficient for production in healthy soil without addition (US EPA 2016b).

Plants require a selection of nutrients to thrive, which are cycled through the environment as part of natural processes. The three main macronutrients involved in plant cycling are the macronutrients nitrogen, phosphorous, and potassium. The most limiting nutrient is nitrogen, due to its specificity of forms, some of which can only be used by plants at specific phases in their lifecycle. Too much of this nutrient can be harmful to the environment, particularly when present in water due to runoff or erosion. Excess nitrogen in the form of nitrate in drinking water results in methemoglobinemia, which is the reduced ability of the body to carry oxygen in the blood. Infants are particularly susceptible to this disease. Nutrient imbalances that affect water bodies can also result in eutrophication and death of aquatic life. Proper nutrient management minimizes this risk and reduces the need for additional fertilizers. An example of nutrient management controls are conservation buffers which enhance infiltration by slowing runoff and uptake nutrients before they reach water bodies.

Assessment tools are an important part of effective nutrient management. Through regular monitoring, farmers can assess soil nutrients and effectively manage plant health. Soil testing and nutrient assessment can also help inform decisions about fertilizing and watering crops to produce the best yields. There are many tests which can be used to monitor soil nutrients. Traditional soil testing monitors nitrogen, phosphorous, potassium, soil pH, organic matter, and electrical conductivity. More specialized testing such as nitrate testing is also available to help monitor this specific macronutrient. Other monitoring such as organic material testing can also be utilized as a nutrient management tool. This type of assessment is most useful in areas where manure, sewage sludge, or waste diversion tactics are used for fertilization on fields, rather than commercial products (EPA 2016).

Construction of a nutrient budget and accounting of nutrient sources are another way in which farmers can track important nutrients and manage resources. A budget requires the farmer to have laid out in advance their target crops as well as have obtained information about their soil character. Good agriculture management helps determine the planting order of crops as well as the most efficient use of the soil resources by planning nutrient usage. In addition to soil testing, nutrients must be quantified by accounting for inputs from outside sources such as commercial fertilizer, manure, atmospheric deposition, irrigation water, and legume credits (legumes put nutrients back into the soil as part of their natural growing process) (EPA 2016). The growth stage of the crop can also be incorporated into the nutrient budget to help determine when and by what method nutrients are best applied. Climate conditions can also affect the transport and transformation of nutrients and can also be made a part of the nutrient budget. For instance, in Arizona, where monsoon rains only occur in the summer months, nutrient application is often delayed until the fall to reduce runoff and erosion of applied nutrients and protect water bodies.

Buffering is a common agriculture management practice that helps retain applied nutrients and conserve existing elements. There are four main types of conservation buffers: riparian buffers, vegetated waterways, filter strips, and vegetated barriers. Each of these conservation techniques performs the same function—to stabilize existing ground by preventing erosion and support productive soil. Riparian buffers are those buffers that are in proximity to a forest and stream or waterway. These woody areas intercept nutrient runoff and sediment while providing habitat to forest creatures. Riparian buffers, when constructed, follow a three-zone concept. Zone 1 is the area closest to the waterway, zone 2 a minimum of 15 feet from the edge of zone 1 to provide infiltration and erosion control from minor overflow, and zone 3, an area 20 feet from the edge of zone 2, to provide major erosion and nutrient washout control (EPA 2016).

Vegetated waterways are another form of conservation buffer used for nutrient management. These grassed areas are either natural or constructed channels that are stable ways to convey runoff and encourage infiltration. Vegetated waterways are typically used as natural pretreatment for fields incorporating filter strips. Filter strips occur along the lower edges of fields and are used to capture outgoing runoff that contains nutrient overflow and erosion. They slow the velocity of water exiting vegetated waterways, allowing suspended soil particles to drop out of solution and thereby reduce nutrient loads in receiving waters. The benefits of this buffer control technique are twofold; reduced erosion and reduced application of nutrients as less are lost to local water bodies. Filter strips also allow for adsorption of pollutants onto plant surfaces and uptake and filtration of soluble pollutants. In addition to these benefits, filter strips also serve as forage for animals and additional biomass for borders, turnrows, and hedges.

Vegetated barriers are also common conservation buffers than slow nutrient loss. This technique controls erosion and traps sediment using stiff grasses which can withstand high water velocities. These grasses disrupt normal water flows and spread out surface runoff, which in turn, improves nutrient and soil retention. Unlike other buffers, vegetated barriers are constructed as berms meant to impede flow rather than channel it or encourage infiltration. They are most commonly used in areas that experience extreme variation in precipitation or irrigation and are impacted by large volumes of water over a short time period. Guidelines for vegetated barriers include density of 50 stems per square foot, a minimum width of 3 feet and use of native grasses whenever possible (EPA 2016).

Water resources management is another important aspect of agricultural practices management. In all aspects of agriculture, water is required— from watering of seeds, to supporting growth of mature plants, and to mixing with soil amendments as fertilizers. In the United States, over 80% of the nation's consumptive water use goes to agriculture (USDA 2008). Effective management of water resources for agriculture is important to reduce waste and protect receiving water bodies. Eutrophication, or nutrient pollution, is an important issue for water resources management of agriculture.

Eutrophication occurs when water runoff into a stream or lake contains a high amount of nutrients or fertilizers. Agriculture is one of the largest contributors to eutrophication, with nutrient pollution coming from animal manure, fertilizers, and even aquaculture. Chemical fertilizers are frequently applied to fields in excess of crop needs, causing an overabundance of nutrients in the soil (MA 2005). Between 1960 and 1990, global use of synthetic nitrogen fertilizer increased seven-fold, while application of phosphorous amendments nearly tripled (MA 2005). This increase in soil nutrients leads to increased eutrophication from water runoff of these over-amended soils. Studies have found that more than 20% of applied nitrogen leaches into groundwater or surface water after application (MA 2005).

When eutrophication occurs, the excess nutrients cause increased growth of microalgae, macroalgae, and phytoplankton. These organisms spread rapidly, causing "red tides" in coastal areas or "pond scum" in enclosed lakes. The algal blooms use up the oxygen in the lake, suffocating other life and covered the lakes' surface, reducing the amount of sunlight and, therefore, photosynthesis of other organisms. This suffocation of a water body is called hypoxia. When hypoxia occurs, dead algae and other organisms sink to the bottom of the water body and are decomposed by bacteria. In the decomposition process, the bacteria use the available dissolved oxygen. Temperature and mixing differences within the water body limit the amount of oxygen that can be replenished, creating stratification. Benthic organisms which are bottom dwelling cannot flee low oxygen stratifications and die, perpetuating the cycle.

Eutrophication can cause a lake, river mouth, or stream to become an ecological "dead zone" almost overnight. One red tide event, which occurred near Hong Kong in 1998, wiped out 90% of the entire stock of Hong Kong's fish farms and resulted in an estimated economic loss of $40 million USD (MA 2005). Nutrient imbalances that affect water bodies can also result in eutrophication and death of aquatic life. Reductions in biodiversity caused by eutrophication can have long-term effects; for instance, damage of coral reefs, as coral reef larvae are disfavoring over algal growth (MA 2005).

Increased agricultural production is partially to blame for eutrophication. Demand for meat (which is one of the largest consumers of nitrogen; it makes up a large percent of elemental material in animal feed) is projected to increase by 14% worldwide by the year 2030 (FAO 2009). This translates to nearly 45.3 kilograms of meat per person, for an estimated increase of nearly 53% when accounting for population growth (FAO 2009). Increased purchasing power, particularly in South Asia, which has a historically low record of meat consumption, is driving the increase in demand, and in directly increasing eutrophication of water bodies. Competing with production of fodder for livestock is the biofuels market, which also is a large user of fertilizers and, therefore, a contributor to eutrophication of water bodies. Biofuels are a burgeoning area of alternative energy, but again, growing these crops requires large amounts of nitrogen and phosphorous amendments to increase yields and profit for agriculture. As a result, fertilizer consumption is expected to increase 40% between 2002 and 2030 in the biofuels division of the agriculture sector (FAO 2009). The majority of the projected increase in global fertilizer consumption is attributed to the developing world where food production and adoption of intensive agricultural practices are expected to increase (FAO 2009). As part of increased food production, land use practices are also changing. In areas that were not previously used for agriculture, removal of native vegetation shifts soil balance and necessitates greater use of fertilizers to replace those natural nutrients when farmed. This additional fertilizer usage in pristine soils (which frequently runs off into receiving water bodies, as soil cannot hold the additional nutrients) results in eutrophication of waters in areas not previously impacted, compounding the problem as farmers seek new lands to keep up increased production.

Climate change is also a driver of eutrophication resulting from agriculture. Managing climate changes can be difficult for farmers as conditions deviate in increasing degrees from previous norms. Increases in temperature means that farmers need to use more water to support crop growth due to increased rates of evapotranspiration. Scarce water resources are also a factor, as sources are not replenished due to drought (particularly in areas that rely heavily on agriculture, such as California). Climate change also means increased use of fertilizers as changing weather patterns mean that nutrient replenishment of plants during growth is perturbed. Increased surface water temperatures linked to climate change could also lead to stratification of the water column, thus preventing oxygenation of the colder bottom waters and leading to more extreme hypoxic or anoxic conditions in systems already suffering from eutrophication.

Warmer water temperatures combined with increased nutrient loads may be beneficial for certain harmful algal species and lead to increased frequency of harmful algal blooms. One example is the abundance of the dinoflagellate *Gambierdiscus Toxicus*, associated with fish poisoning, which has been positively correlated to increased surface water temperatures in tropical oceans caused by the meteorological phenomenon El Niño (Moore et al. 2008). Changes in precipitation patterns linked to climate change can also influence the expression of eutrophication. For instance, an increase in precipitation may lead to changes in stratification patterns as more freshwater in the form of runoff (and associated desalinization) is discharged into coastal oceans, and will also translate into higher nutrient fluxes (Rabalais et al. 2008).

3.1   SLASH AND BURN FARMING IMPACTS AND CONTROLS

Slash and burn agriculture is a destructive form of agriculture that is most widely used in rainforest areas, such as the Amazon. In this method of farming, cover vegetation such as trees and forest undergrowth is clear cut, and the resulting vegetation is burned to provide fertilizer and dispose of the covering plants. The land is then planted with the crop that is desired. As this method strips the land of nutrients by removing native vegetation, the area is only suitable for crops for a short time, sometimes as little as two to three years. The soil also loses valuable nutrients through erosion during the rainy season, as the covering root structure has been removed and the soil can then wash away. Nutrients are drained from the topsoil during the rainy season due to the lack of canopy cover and are baked off during the dry season for the same reason.

Farmers try to lengthen the growing period in an area by applying soil amendments such as fertilizers and pesticides once the soil is exhausted. As the fertile topsoil layer has been eroded or overplanted, these measures are short-lived. Applied fertilizers and pesticides leach off of the amended land during the rainy season and pollute local waterways. During dry seasons, dust containing pesticides are carried by the winds to affect pristine areas.

In areas that are less populated, slash and burn farming can be sustainable, as the land has time to recover and regrow between periods of burning. This buffer must be at least 5 years, ideally 10. As areas of the rainforests become more populated due to migration of native tribes today, this interval is not usually reached and the land remains infertile after use. In areas of rich vegetation and diverse habitat, this loss is irreparable.

The Amazon is particularly vulnerable to slash and burn agricultural practices, as the richness of the environment is largely due to living flora and fauna rather than thick nutrient-rich soils. In fact, the soil of the Amazon is a thin layer of biomass that is continually replenished by the living organisms that inhabit it. When these organisms are gone, burned out by fire, the soil is no longer replenished and quickly loses its fertility. The Amazon is also defenseless against fire, having plants that are not suited to burning and have no natural fire immunity (Figure 3.1). In fact, experts

Figure 3.1. Before and after. An 1843 oil painting of Brazil's Atlantic rainforest, and a recent burning event in the same forest.

Source: G. L. Peixoto/ ICMBio, Brazil: (painting, inset) Photographer: J. Acioli; Felix Emile Taunay/Museu Nacional de Belas Artes/IBRAM/MinC.

estimate that natural fires in the Amazon have occurred on the order of hundreds of years as lightning is typically accompanied by rainfall.

In some regions, slash and burn agriculture is followed by cattle ranching on the fallow fields. The infertile soil is compacted by cattle grazing which encourages shallow root systems such as grasses and woody forest growth and must be reburned frequently to encourage this growth to occur (Figure 3.2). The practice may be even more harmful to the rainforest as native plants are crowded out by nonnative grasses and the landscape does not have a chance to replenish needed soil nutrients.

Another consequence of slash and burn agriculture is the leaching of black carbon into native soils and the marine ecosystem from centuries of unsustainable farming practices. In the Atlantic Forest in Brazil, nearly 40 years after slash and burn agriculture has been banned, black carbon from this practice was still being leached into the nearby waterway of Paraiba do Sul River, the largest river that exclusively drains the area, as late as 2012 (Dittmar et al. 2012). It was calculated by researchers that over the course of slash and burn agriculture, 200 million to 500 million tonnes of black carbon were released. Calculating using the half-life of black carbon, it will take between 630 and 2200 years for just half of that mass to leach out of the region's soils. When that mass leaves the soil, it leaches into waterways like the Paraiba do Sul, contaminating the river

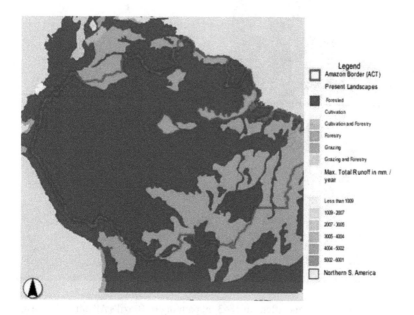

Figure 3.2. Cattle ranching and cultivation within the Amazon.
Source: American Geosciences Institute.

and later the ocean (Dittmar et al. 2012). When black carbon is leached from the soil into the river by rain, it is carried downstream to the ocean, and later to deposits on the deep ocean floor. Black carbon deposits from tropical deforestation have been found in ocean sediments all around the globe, including Antarctica. The impacts of tropical deforestation are widespread as shown by these data.

One method of control of slash and burn agriculture is the encouragement of sustainable forestry techniques and conservation agriculture. Farmers need to be educated on the practices of alley-cropping (co-planting food crops with native trees), creation of buffer zones of native plants around existing forests to prevent incursion of nonnative species, planting of a diversity of crops to prevent depletion of nutrients, and reforestation techniques.

Rights to resources in an area are also typically poorly understood, either inherited or taken as rights of occupation or usage. Allocation of forest areas in this way encourages premature harvesting and clearance of the land, which in turn contributes to population migration and ultimately climate change. Agricultural expansion driven by rising food prices or population growth encourages the adoption of unsustainable agriculture practices such as slash and burn farming. By better understanding the social and economic underpinnings of such practices, better controls can then be put in place. Other controls that can be used to reduce slash and burn farming practices include incentives to retain land as forested and the adoption of trade suited to the region, such as the production of native honey or chilies by tribes where these resources are naturally abundant.

It is also important to remember that rainforests are enormous carbon sinks. When these lands are cleared, this carbon is released to the atmosphere, contributing to global climate change. In a 2007 article in Science, Renton Righelato and Dominic Spracklen calculated the amount of carbon released to the atmosphere when forest is cleared to be nine times more than the amount that can be recouped by planting biofuels in the same area (Righelato and Spracklen 2007). Keeping rainforest pristine through protection and sustainable agriculture practices preserves biodiversity and reduces nutrient runoff and soil erosion.

3.2   OVERWATERING AND RIPARIAN RIGHTS ISSUES

Riparian water rights or simply riparian rights are a system of allocating water rights to those who own land along waterways, with its roots in English common law (Guerin 2003). All landowners who possess land adjacent to a waterway have the right to make use of its flow through or over their property. If the land owner requires more water than flows through their property, rights are distributed in proportion to the frontage of the land with respect to the water source. Rights are attached to the property and cannot be sold or transferred (Guerin 2003).

Riparian rights include such things as the right to access for swimming, boating, and fishing; the right to erect structures such as docks, piers, and boat lifts; the right to use the water for domestic purposes; the right to exclusive use if the waterbody is non-navigable. All riparian rights are governed by the criteria of "reasonable use" which state that downstream owners are entitled to receive water that is undiminished in flow or quality (Guerin 2003). The "natural flow" doctrine is also part of riparian rights and covers the flow of the river through or around the property such that for the following users the flow of the river continues in its natural state, undiminished by consumption or pollution (Ausness 1978). In the case of owners on either side of a waterway, the riparian rights of each extend to the center of the waterway and no further, as shown in the Figure 3.3. Owners are responsible for maintaining the trees, shrubs, and fences that border the waterway as well as clearing any obstructions on their land that may impede flow and adversely affect the rights of those downstream.

In the eastern United States, riparian rights are the governing legal doctrine, while prior appropriation is generally the rule in the western states. Prior appropriation means simply that the body "first in time is first in right," or that those with the earliest claim upon the water have the greatest rights associated with it and those following are only granted their allocation once that request has been fulfilled. A good example of this is the Central Arizona Project which brings water from the Colorado River into Arizona for use in the region. Due to prior apportionment, the water from the Colorado River within the Central Arizona Project is subjected to stifling in times of drought as rights are subordinate to California, which has an earlier claim upon the waters of the river (Law of the River, US

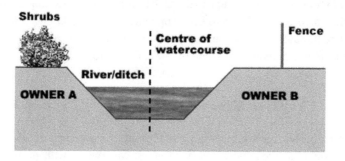

Figure 3.3. Illustration of riparian water rights.
Source: Eye on Calderdale, 2015.

Bureau of Reclamation 2008). Apportioned water rights are derived from beneficial use of the water rather than frontage upon the waterway. Use and conveyance of the water much be efficient and are usually described as a quantity, such as cubic feet per second for diversion or acre feet for storage.

Prior appropriation has been challenged recently in the courts by native tribes that seeking recognition as first users of the water, even if they were not the first permitted users of the water (NPR Water Wars 2013). On the Klamath River in Oregon, irrigators using the water upstream of the Klamath Indian Tribe were told to stop or greatly reduce their water use so there is enough water in the tributary for the native fishery. The state of Oregon ruled that the water rights of the tribe superseded any permits for users on the tributary, as they had the oldest claim to the water (NPR Water Wars 2013). The Klamath Tribes venerate the native suckerfish that is now on the endangered species list due to lack of water from the Klamath River tributary. By shutting off irrigation from farmers upstream due to prior appropriation rights, the tribe is hoping that the fish will become plentiful once again.

Overwatering is overreach of water rights by an owner that falls outside of reasonable use or natural flow doctrine. Users that exceed their allotments of water are subject to fines and legal action within the United States, but this action usually takes many years of court battles. Overuse of water can take the form of wasteful use upstream that deprives users

Figure 3.4. Don Gentry, chairman of the Klamath Tribes, says the tribes have not been able to fish for suckerfish for past 27 years. "The condition of our fish is just so dire," he says.

Source: Amelia Templeton for NPR, Water Wars, 2013.

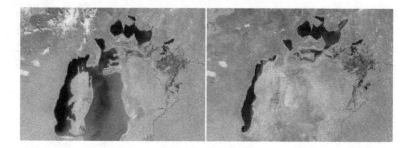

Figure 3.5. In 2000 (left), Asia's Aral Sea had already shrunk to a fraction of its 1960 extent (black line). Further irrigation and dry conditions in 2014 (right) caused the sea's eastern lobe to completely dry up for the first time in 600 years. *Source:* NASA Earth Observatory, 2014.

downstream, or it can mean that groundwater is used at a rate that is much greater than its rate of replenishment from precipitation (Crops and Drops 2005). One of the most striking images of overuse by agriculture is that of the Amu Darya River which feeds the Aral Sea. The Amu Darya has been entirely used up by agriculture for irrigation of cotton plantations.

The Amu Darya River is not the only river that has been deprived of its flow by agriculture. The Yellow River in China is also overused and does not reach the sea for over seven months of the year as farmers consume its water for irrigation (Crops and Drops 2005). In the United States, the Colorado River has the same problem, and only now reaches its end in Mexico due to increased controls on agriculture within the United States. The overuse of groundwater for agriculture has serious implications for food production and will continue to be an issue as global climate change induces changes in precipitation and temperature around the globe.

CHAPTER 4

Resource Extraction Issues

4.1 OPEN-PIT MINING

An open-pit mine is an extraction or cut made at the surface of the ground for the purpose of extracting ore from the Earth's crust. The mine remains open at the surface for the duration of its life. Minerals that are commonly mined using this technique include copper, gold, iron, and aluminum (Mine-engineer 2012). In the extraction of ore from the mine, layers of rock are exposed; the ore separated from the nonvaluable or waste rock; and then the next layer of ore or waste rock is removed. These layers of rock are called benches and several may be in operation simultaneously within an open-pit mine (shown in Figure 4.1). (Basics of an open pit mine 2012).

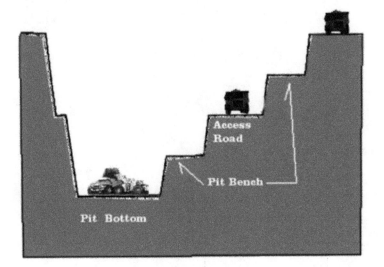

Figure 4.1. Simple schematic of an open-pit mine.
Source: Basics of an Open Pit Mine, 2012.

While open-pit mining is one of the most common ways to mine resource minerals, it is also one of the most unsightly and environmentally damaging methods of mining. Open-pit mines can reach miles long and deep, creating large scars on the Earth for what is often small amounts of minerals extracted. The technique of open-pit mining goes back over 100 years, and there are mines that have been in continuous operation for the history of open-pit mining. The Hull Rust open-pit mine in Hibbing, Minnesota, has been in operation since 1895 and mines iron ore. It is one of the largest open-pit iron ore mines in the world at 5-mile long, 2-mile wide, and 535-feet deep (LosApos 2017). The largest open-pit mine in the world is in Garzweiler, Germany, where lignite is mined (Figure 4.2). The mine is over 18.5 square miles and several villages have been moved to accommodate the mining operation. When the mine is exhausted, it will be filled with water to create Germany's second largest lake (LosApos 2017).

By their nature, surface mining operations alter the Earth's surface. Not only are rock and minerals removed from the Earth, but also before mining can begin, vegetation, soil, and trees must be removed to allow access to the ore. The topographic modifications take place on a large scale which causes significant environmental damage. Sites on the mine must be created for the disposal of waste rock or overburden, access roads must be built for heavy equipment (which is also damaging), as well as mineral processing, disposal, and extraction areas (Footprints in the dust 2017). The processed rock or tailings must also be stored at the mine, resulting in settling ponds and contamination of surface and groundwater.

Figure 4.2. Strip mining machines working at the Garzweiler mine.
Source: LosApos, 2017.

The individuals working at the mine will build structures of offices, workshops, and laboratory facilities to assist the operations of the mine, further expanding the area that is impacted by the mining operation (Footprints in the dust 2017). One of the hallmarks of an open-pit mine is a tailings pond or pit that stores the water-borne waste of processed ore. These embankments are not meant to be permanent and are usually constructed of compacted Earth, which can fail catastrophically in times of extreme precipitation or weather. In one such instance, an tailings pond dam at the Mount Polley open-pit gold and copper mine failed in 2014, releasing more than 10 million cubic meters of contaminated mine slurry into a local waterway (Footprints in the dust 2017). An iron ore containment damn at the Samarco mine in Brazil failed in 2015, releasing 81 million cubic yards in mining waste into the Rio Doce, one of Brazil's most important rivers. The dam failure destroyed over 600 homes in a nearby village, which claimed 20 lives and slurry is still contaminating beaches in Rio De Janeiro and Bahia (Figure 4.3).

Open-pit mines also affect the qualities of soil and groundwater, contaminating these natural resources for years after mining has ceased. In the case of soil, the acidity and salt content are frequently altered from exposure to the harsh chemicals that are used to separate the ore from the waste rock. Particulates from metal smelters settle within the dust that is adjacent to the mine, affecting soil quality and human health. Erosion of soils is also an important environmental consequence of open-pit mining. Stable soils are disturbed by

Figure 4.3. Image of the Samarco Mining disaster.
Source: Marcelo Silva, youtube.

mining activities, causing erosion from water and wind. These soils are then deposited in local or regional waterways, impacting aquatic habitat (Footprints in the dust 2017). Contamination of soils by toxins is also an important environmental impact of open-pit mining. Chemicals used in processing of ores can result in nutrient imbalances which affect the regrowth of plants after mining has stopped, as well as the soil profile of an area. Alteration of the soils profile prevents native plants from recolonizing a mined area as well as making it unsuitable for other usage after mining has ceased.

Changes in water quality are also an important environmental impact of open-pit mining. Large amounts of water are used in processing of ores to separate minerals from waste rock. The water is often poorly treated and stored in unstable earthen dams and pits that surround the mine. The problem is known as acid mine drainage and is a serious issue associated with coal mining in the eastern United States. Drainage from mines impacts both surface and groundwater and can often take decades to remediate. In the 1990s, the U.S. Forest Service estimated that between 20,000 and 50,000 mines on federal lands were generating toxic acid discharges that adversely affected between 5,000 and 10,000 miles of streams (Footprints in the dust 2017). One recent news story is that of the Gold King acid mine drainage discharge into the Animas River in Colorado in August of 2015 (Figure 4.4). The mine released some 3 million gallons of acid drainage into the river initially and continues to discharge nearly 600 gallons of contaminated water into the river daily, turning the waterway to a bright yellow color (Footprints in the dust 2017). The wastewater from the mining operation contains lead, arsenic, cadmium and other chemicals that are left over from ore separation activities when the mine was in operation. This water was never remediated once the mine had closed and the spill was triggered by accidental removal of a dam of rock.

Figure 4.4. The Animas River flows through the center of Durago on August 7, 2015, following the failure of a dam associated with the Gold King mine.
Source: Denver Post, 2015.

Groundwater quality is also an environmental concern related to open-pit mining. Mines use a tremendous amount of water for operation and open-pit mines often intersect with the groundwater table, impacting local aquifers. In Canada, 4% of the total yearly water usage comes from mining, while in Australia, 2% to 3% of yearly water use is due to mining operations. In the United States, 1% of yearly water use in industry is for mining; about 5.32 billion gallons per day (Maupin et al. 2014). The majority of water used for mining in the United States is groundwater, and water usage for mining operations increased in usage from 2005 to 2010 (Maupin et al. 2014).

Groundwater must be pumped out of the mine, and once the mine closes, the water is often left to accumulate at the base of the mine pit, creating contaminated lakes which can then leach into other waters. During mining operation, once the mine level intersects the water table, dewatering must occur. Dewatering occurs through boreholes that are drilled into the rock adjacent to the mine (Australian Government 2014). Water is drawn up through these boreholes like water through a straw to lower the water within the mining pit, and this water is then deposited on the surface in holding ponds or pools (Australian Government 2014). Water that is not removed with dewatering boreholes and collects in the lowest level of the mine must also be removed and this is called interception. Interception typically uses pumps that move the water upward out of the mine, again depositing it in holding ponds or pools. Removal of groundwater from the aquifer to accommodate mining operations can have a significant effect on the local water table, reducing or eliminating access to waters downstream (Figure 4.5).

In addition to affecting the availability of pristine water from the aquifer, water moved in mining operation is typically contaminated by detritus from the mine and is no longer pristine. This water must be treated on site before being released into local waterways.

Water usage in mining operations is also significant. There are four categories of water usage that occur during mining operation: mill water,

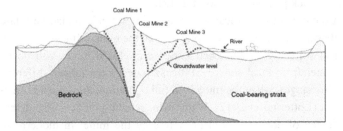

Figure 4.5. Conceptual cross section of aquifer dewatering from groundwater drawdown.

Source: Australian Government, 2014.

process water, leachate, and effluent (Lottermoser 2012). Any other water that comes into contact with mining operations is generically referred to as mining water. Mill water is water that is used in the mineral extraction process. Water used in this operation crushes and grinds ore and typically contains dissolved minerals or metals (Lottermoser 2012). Mill water is followed by process water, which is the water used in the chemical extraction of minerals. This water is typically highly contaminated with chemicals such as acids or abrasives that are used in mineral extraction (Lottermoser 2012). Leachate is water that is derived from solid mine wastes such as waste rocks or tailings. It may contain dissolved minerals, process chemicals, and other mining residues. These mine waters are collected and stored in tailings ponds before being treated and released as effluent (Lottermoser 2012). The extent of contamination of water by mine water varies with respect to the type of ore being mined, extraction process, climate, stage of the mine, and management practices in place. Ores containing sulfides are particularly reactive and can easily contaminate water by forming sulfuric acid, while other ores are not as reactive and do not require as extensive treatment. Common contaminants in metal ores include metals, metalloids, and salts (Lottermoser 2012).

In the ore extraction process, a variety of chemicals are used that can be harmful to water. Harsh organics such as sulfuric acid and cyanide are common process chemicals that are used in mining. These chemicals are used in both physical and chemical process streams to separate ore from waste rock and generate highly contaminated water (Lottermoser 2012). Contaminated water is collected in tailing ponds prior to treatment and release as effluent.

Climate is also a factor in water contamination. In arid areas where mining is common, groundwater must be mined in order to support operations, which can adversely affect the local water table. Conversely, in wet areas, overflow of tailings ponds during times of high precipitation is possible, as well as leaching of contaminated water from tailing mounds or waste rock piles (Lottermoser 2012).

Contamination of local waters by mine water can also be related to the life stage and management practices of the mine. Mines in early stages use less water than established mines, and closed mines will no longer take water from local sources. Process water is the largest difference in mine life stages; when the mine is in full operation, this expenditure is the greatest (Lottermoser 2012). Water management practices also determine the degree of contamination of waters by the mine. In modern mines, significant practices and protocols must be followed in accordance with current environmental laws, while old mines were not in operation under those rules and have greater environmental consequences.

For example, artisanal mining, or ASM, for gold poses a serious threat to water quality in areas where it is being done (Lottermoser 2012). ASM mining uses liquid mercury as part of the ore extraction process, and accounts for nearly one-third of all human released mercury (Telmer 2006). While ASM only accounts for 20% to 30% of gold mining worldwide, releases of mercury from this operation can be as large are 1,000 tonnes per year (Telmer 2006). ASM operations are small and widely scattered, so it can be difficult to enforce environmental management practices and to sanction polluting operations.

Other operations, such as coal or aluminum extraction, while no less harmful, are typically much larger and better regulated. In the United States, mining discharges are regulated under the Clean Water Act, Toxic Substances Control Act, National Environmental Policy Act, Resource Conservation and Recovery Act, and the Comprehensive Environmental Response, Conservation, Recovery, and Liability Act (American Geosciences Institute 2017). Within Canada, the law is much more narrower and mining discharges are regulated under the Mining Metal Effluent Regulations as part of the Fisheries Act (Natural Resources Canada 2010). In over 100 countries, environmental impact statements (EIAs) are required for mining operations. EIAs outline the environmental management strategy of the mine as well as identifying potential sources of water contamination and strategies to mitigate these effects.

One of the most significant areas of water contamination from mining is tailing ponds. Tailing ponds are holding areas for the detritus of mining operations, called tailings. Tailings are made up of finely ground rock and minerals from mineral processing (Younger et al. 2002). This waste rock is often generated as a water-based slurry, which is kept on the mine site in sedimentation ponds. Tailings ponds prevent the direct discharge of mining water waste into surface waters. In dry climates, the water is simply allowed to evaporate from the tailings slurry, while in wet climates, tailings slurry can be treated and then released as effluent when the water meets acceptable standards (Younger et al. 2002). Tailing ponds are temporary structures and have been known to fail when improperly constructed, such as in the case of Baia Mare.

Baia Mare is a historic incident in which a tailings ponds dam failed at the Baia Mare Aurul gold mine in Romania in 2000. When the tailings pond dam failed, this released over 100,000 cubic meters of mining wastewater into the tributaries of the river Tisza in Hungary (ReliefWeb 2000). The spill resulted in extensive fish kills due to the high levels of cyanide in the wastewater. Other native fauna such as mute swans and black cormorants were also affected by the toxic water release (ReliefWeb 2000). This accident generated the adoption of the International Cyanide

Management Code, which focuses on the safe management of cyanide in mining operations. Companies adopting the code are audited for their cyanide management practices and receive a seal to show their compliance (International Cyanide Management Institute 2012).

Another area of water contamination is heap leaching. In the practice of heap leaching, crushed ore is piled into a mound, and then sprayed with process chemicals to separate the minerals from the waste rock (Figure 4.6).

Heap leaching is used for a variety of ores, including copper, uranium, and gold (Petersen and Dixon 2002). The crushed ore is comingled, piled onto an impermeable liner, and then the process chemicals are applied to the heap. Chemicals are either sprayed using a sprinkler system or the rock is irrigated with a drip system. The drip system is preferred during modern operation as it minimizes evaporation and distributes the chemicals more uniformly than sprinkler systems (Petersen and Dixon 2002). The chemicals percolate through the heap to separate the target minerals from the ore. The process is a slow one, with the quickest leach cycle taking one to two months for simple oxide ores like gold ore and upward of two years for complex ores such as nickel laterite (Petersen and Dixon 2002). The used leach solution is captured and collected to be recirculated through the heap, once additional chemicals have been added. At the same time, the dissolved minerals are removed and captured. Leachate that is exhausted is treated as mine wastewater and then released as effluent or kept in holding ponds (Petersen and Dixon 2002). Leachate wastewater is

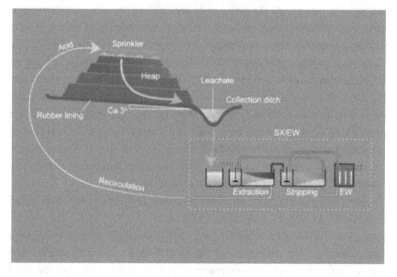

Figure 4.6. Illustration of heap leaching followed by solvent extraction and electrowinning.

Source: BioMineWiki, 2007.

very toxic, containing high concentrations of dissolved solids, metal salts, and acid residue. Leachate allows for mobilization and bioavailability of metals from rock that can then negatively affect the surrounding sediments and plants and animals. Bioavailable heavy metals retard growth in plants, negatively affect the development of biota, and can decimate the aquatic organism population of streams (Petersen and Dixon 2002). Leachate can also be acidic, and can result in acid mine drainage if not properly contained. Acid mine drainage is harmful to benthic microorganisms which populate stream beds and support healthy waterways.

Degraded air quality is a significant issue associated with open-pit mining. Open-pit mines emit gas and particulates to the surrounding atmosphere that negatively impact air quality. These emissions come from a variety of sources associated with the mining process including blasting, crushing, loading, and transferring of rocks and ore (Environmental Code of Practice for Metal Mines 2017). Tailings piles, waste rock piles, and open-pit benches are all sources of wind-blown particulate matter that has the potential to negatively affect environmental and human health.

Smelting of rock to separate minerals from ore is a source of environmental air contamination (Environmental Code of Practice for Metal Mines 2017). The most common air contaminant from the smelting process is sulfur dioxide. Sulfur dioxide reacts with atmospheric water vapor to form sulfuric acid or acid rain (American Geosciences Institute 2017). Acid rain percolates into soils and sediments, harming existing vegetation and retarding the growth of new plants. Barren ground areas near smelters are common, and are due to the locally degraded environment from mining (American Geosciences Institute 2017). Air contamination in the form of dust and acid rain can also endure beyond the lifetime of the mine, creating legacy environmental effects. Some mines do try to control air quality degradation by spraying water on tailings piles to minimize dust, revegetating tailings piles to minimize erosion and blown particulate matter, and treating gas emissions from smelters (American Geosciences Institute 2017).

The terrestrial environment is a concern when resources are extracted for open-pit mining. As part of the mining process is carving tiers and shelves into the bedrock, flora and fauna are removed from the surface to accommodate the process. The stripping of the surface during mining and exploration can have significant effects on plant communities. In areas that have been mined, loss of vegetation and natural habitat allow for invasion of alien species and nonnative plants which may retard the growth of native species. These disturbances in landscape in turn alter soil strata and invertebrate species. Loss of plant communities also translates to loss of wildlife breeding grounds and migration corridors (Environmental Code of Practice for Metal Mines 2017).

Mining activities may contaminate existing and new plants through the uptake of minerals from contaminated soils. In heavily contaminated areas, vegetation may be stunted or unable to take root entirely (Environmental Code of Practice for Metal Mines 2017). As part of restoration and reclamation, mines are required to remediate any impacted areas. In the United States, this falls under the Surface Mining Control and Reclamation Act (SMCRA), which was established in 1977 for regulation and reclamation of coal mines. The SMCRA states that mining operations shall establish "a diverse, effective, and permanent vegetative cover of the same seasonal variety and native to the area . . . and capable of self-regeneration and plant succession . . ." [Section 515(b)19] (Surface Mining Control and Reclamation Act, Public Law 95–87 Federal Register 3 Aug 1977, 445–532). Introduced species are allowed under SMCRA, but only where necessary to achieve the postmining land use. In Appalachia, where most open-pit coal mining takes place, revegetation is frequently accomplished with fast-growing grass and legume species, most of which are nonnative (Figure 4.7). Tree cover is also required as part of SMCRA, and many operators choose the Eastern white pine, which is native to the southeastern United States as it is commercially valuable (Holl 2002).

In a Virginia study of reclaimed mine lands versus pristine forest, it was found that revegetated sites were lacking in species diversity provided by native plants. Specifically, reclaimed sites had only 54% as many herbaceous and

Figure 4.7. Progressive revegetation of a mining site.
Source: Polster, 2015.

62% as many woody species as unmined forest (Holl 2002). Recommended strategies to encourage species diversity in reclamation include greater diversity of trees in replanting, spreading topsoil from newly mined areas on areas being reclaimed, and providing forest buffer near mining areas to provide sources of native seeds that are not easily dispersed (Holl 2002).

There are a number of hurdles to revegetation of mined areas, particularly open-pit mines. It is common to plant dense covers of grasses and legumes, but these plants can prevent woody species growth. Forest tree cover planted in compacted, low nutrient waste rock generally fails to thrive as well due to a number of factors (Polster 2015). In many mine sites, the land is severely compacted from years of use by heavy equipment and simply disturbing the area with a large bulldozer does not adequately address compaction and fails to create the micro-site diversity needed for vegetation establishment. Additionally, most waste dumps are at severe angles and will not revegetate naturally. Similarly, the size sorting of materials on waste dumps means that coarse-textured materials end up on the lower slopes (Polster 2015). The lack of moisture holding capacity with these materials means that vegetation establishment is limited. Mining wastes are generally lacking nutrients that are essential for plant growth. In some cases, chemical extremes (e.g., acid rock drainage) also prevent vegetation establishment and growth. Nonnative, invasive species can overwhelm recovery processes, as can excessive predation by herbivores, such as deer, which prevents adequate regrowth and establishment of native species (Polster 2015).

Compaction is probably the most common difficulty found at mining revegetation sites. Spreading materials with bulldozers creates compaction in the soils and prevents healthy growth of vegetation. Using an excavator to prepare the surface of the compacted area can be an effective way of alleviating compaction which encourages effective revegetation of areas (Polster 2015). Using the excavator, operators can roughen mining surfaces and loosing compacted Earth to provide suitable sites for vegetation reestablishment (Polster 2015). This technique can provide an additional advantage by recontouring slopes to control erosion and provide suitable sites for vegetation establishment.

Steep slopes are common at many mines. Waste rock dumps are constructed by dumping wastes around or over the rim of the dumps and then pushing these materials over the edge of the dump slope, forming steep slopes (Polster 2015). These slopes tend to revegetate very slowly as most of the material on the slope is too coarse to support plants. The materials at the top of the slope are fine textured and will support plant growth, but this area is continually bombarded by materials from above. In the middle of the slope, the materials are too coarse to hold moisture and therefore will not support vegetation. At the bottom of the slopes, the spaces between the large rocks slowly fill with organic matter and eventually will support the growth of higher plants (Polster 2015). This process can be greatly expedited at mine waste dumps by pushing organic materials from the top of the slope over the face and down to the bottom of the slope.

The low nutrient status of most mining wastes can also limit the growth of some plants. However, pioneering species, including those that are associated with nitrogen-fixing bacteria, have evolved to grow in areas of low nutrients, and these species can be used as initial plantings to encourage the establishment of hardier species (Polster 2015). The use of early pioneering species as the principle revegetation species on mining sites eliminates the need for fertilizers while providing a vegetation cover that will build healthy soils on the relatively inert materials left from mining. Acid rock drainage and metal leaching are a problem at some mines that require that cover systems be established. Adverse chemical properties of waste materials (tailings and waste rock) at some mines require that these materials be isolated from the environment. The key to the design of waste rock covers is to ensure that there is ample clean material on top of the material that is used to seal the wastes to allow vegetation to grow freely (Polster 2015).

There is an additional advantage of this technique in phytoremediation. In this strategy, growing plants mitigate adverse chemicals by uptake and return needed carbon and nutrients to the soil to support successional growth of additional plants. One of the overarching issues with revegetation of mining sites is competition by seeded agronomic grasses and legumes that can restrict woody species growth (Polster 2015). Seeded grasses and legumes have historically been part of the mine reclamation tool kit. Making sites rough and loose, adding woody debris, and seeding or planting in pioneering woody species avoid the problems of seeded agronomic grasses and legumes and encourage better reclamation of sites.

4.2 DEFORESTATION

Deforestation has become a significant issue in resource extraction. Deforestation most often results from anthropologic means. Forests are cleared for agriculture or human expansion. The most environmentally troubling areas of deforestation are the clearing of tropical rainforests in South America and Africa.

Over 30% of the planet is forest area, but an area equal to the size of England is cleared each year (National Geographic 2009). Trees are removed to provide land for crops or the grazing of livestock, using a technique known as slash and burn farming. This technique clears land by clear-cutting trees and foliage, then setting fire to the land to burn out the remaining plant material. The land is then planted with crops or used as grazing area for cattle or other livestock (National Geographic 2009).

Logging is another way in which deforestation takes place. Trees are cut to provide wood, pulp, and paper, including the furniture industry. Some forests are sustainably managed, while other logging takes place illegally. Those loggers working outside the law do not replace trees that have been harvested, causing deforestation (National Geographic 2009). Illegal logging companies or individuals build roads to access deeper and older growth forests, leaving a greater impact on the Earth.

Deforestation and irresponsible forestry are also contributors to global climate change. Trees provide not only food and habitat for animals, but also act as carbon sinks and water recyclers in the natural world. Reducing forest leads to ecosystem imbalances in these important natural processes. The World Resources Institute has estimated that between 12% and 17% of global greenhouse gas emissions are contributed annually by deforestation (WRI 2015).

Deforestation also affects land temperature. In areas of high altitude, deforestation causes lower temperatures, as trees are replaced by flatland which accumulates snow, reflecting sunlight (Witman 2017). The tropical low latitudes experience the opposite effect, reducing evapotranspiration and causing warming (Witman 2017). In many areas of the world experiencing deforestation, NASA data from global satellites and flux towers show a diurnal asymmetry, meaning that these effects are not constant with day and night (Figure 4.8).

Figure 4.8. Flux Tower measuring temperature.
Source: Witman, 2017.

Instead, researchers found that deforestation leads to daytime warming and nighttime cooling (Witman 2017). During the day, the land absorbs radiation and heats, while at night trees release stored heat, leading to localized warm areas within the forest (Witman 2017). These changes in temperature flux produce negative effects on local flora and fauna, which must either adapt to the changes in temperature or succumb to the effects.

Other causes of deforestation are the expansion of infrastructure, such as clearing of land for roads or dwellings as part of urbanization (Global Forest Resources Assessment Report 2005). Road building indirectly causes deforestation as avenues into previously inaccessible or difficult to access forest areas can now be harvested and sold. The unclaimed land, once it has been logged, is then taken by settlers for agriculture such as planting crops or grazing, and the loggers move on to other areas. The forest is rarely regained (Global Forest Resources Assessment Report 2005).

Other causes of deforestation include fire, as areas of slash and burn farming adjacent to undisturbed forest may catch fire and be removed as well. Commercial industry also plays a role in deforestation. In the Amazon, soybean farms for large-scale production are a leading cause of deforestation, while in Borneo and Sumatra forests are cleared or converted for the production of palm oil and coffee (Global Forest Resources Assessment Report 2005).

Deforestation follows a distinctive pattern called the forest transition curve, in which the poor economy is boosted by the harvesting of forests, and then an attempt is made to reduce the economic loss once the area has been overharvested. In this transition curve, some countries are recovering faster than others, including India and Costa Rica. In these areas, the deforestation was not as severe initially and the countries are recovering by replanting trees. In 1980, India had about 640,000 sq km of forest left. Now, it has 680,000 sq km, and is replanting about 1,450 sq km a year (Figure 4.9) (The Economist 2014). Costa Rica and Mexico are improving and replanting their forests by implementing a policy called "payment for ecosystem services." This policy reduces deforestation by making the trees worth more standing that they are cut down; the users of clean water and shelter derived from the trees pay for the benefits. The Mexican government paid around $500 million dollars to forest organizations from 2003 to 2011, which then used the funds to sustain the forest area and promote sustainable management practices (The Economist 2014).

Sustainably managed forests plant new trees to replace those trees that have been harvested, minimizing deforestation. Examples of companies that practice sustainable forestry include Weyerhaeuser, Georgia Pacific, and Drummond Press. There are other management practices and criteria adopted by those practicing sustainable forestry that minimize

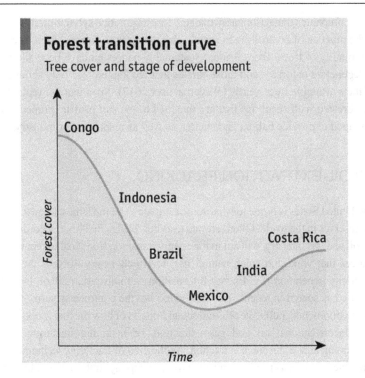

Forest transition curve
Tree cover and stage of development

Figure 4.9. Forest transition curve.
Source: The Economist, 2014.

deforestation. These practices include maintaining biodiversity, sustaining the forest ecosystem such that the forest continues to provide food, pulp, biomass, and wood, the forest remaining part of the water cycle, including carbon capture and preventing soil erosion, providing habitat and shelter for people and wildlife, and offering spiritual and recreational benefit (PEFC 2017). Forests that have been certified as sustainable are accredited by third-party certifiers such as PEFC. The Programme for the Endorsement of Forest Certification (PEFC) is an international nonprofit, nongovernmental organization dedicated to promoting Sustainable Forest Management (SFM) through independent third-party certification.

PEFC works throughout the entire forest supply chain to promote good practice in the forest and to ensure that timber and nontimber forest products are produced with respect for the highest ecological, social, and ethical standards (PEFC 2017). PEFC is an umbrella organization. It works by endorsing national forest certification systems developed through multistakeholder processes and tailored to local priorities and conditions. PEFC oversees not only the forest itself, but also the production and labor practices which are used, including monitoring chemical use and working conditions (PEFC 2017).

Companies that sustainably manage forests practice sylvicultural, which is the practice of controlling the establishment, growth, composition, health, and quality of trees. Trees that are sustainably harvested have been allowed to regenerate naturally, and those forests that are logged lose only about 2% of their available trees yearly (Weyerhaeuser 2017). Sites that are replanted are forested with seedlings that are matched to the soil profile, pruned, and managed to provide habitat and shelter, as well as resources for industry.

4.3 OIL EXTRACTION/FRACKING

The United States is home to what some estimate to be the largest known shale gas reserves in the world. Often referred to as the "bridge fuel" that, according to the oil and gas industry, will aid in the energy transition from coal to renewable sources like wind and solar, natural gas now fuels nearly 40% of the U.S. electricity generation (US Energy Information Administration 2016). Natural gas use has soared in recent years, but so too has the controversy surrounding the environmental, public health, and social impacts of how the fuel is obtained.

Unconventional oil and gas extraction, or hydraulic fracturing, is a burgeoning area of resource extraction. Application of fracturing techniques to stimulate oil and gas production began to grow rapidly in the 1950s, although experimentation dates back to the 19th century. Starting in the mid-1970s, a partnership of private operators, the U.S. Department of Energy (DOE) and predecessor agencies, and the Gas Research Institute (GRI), endeavored to develop technologies for the commercial production of natural gas from the relatively shallow Devonian (Huron) shale in the eastern United States. This partnership helped foster technologies that eventually became crucial to the production of natural gas from shale rock, including horizontal wells, multistage fracturing, and slick-water fracturing (US Department of Energy 2011)

The practical application of horizontal drilling to oil production began in the early 1980s, by which time the advent of improved downhole drilling motors and the invention of other necessary supporting equipment, materials, and technologies (particularly, downhole telemetry equipment) had brought some applications within the realm of commercial viability (US Energy Information Administration 2016). This method of resource extraction combines a new form of horizontal drilling with hydraulic fracturing—more commonly known as fracking. The process blasts open fissures in underground shale-rock formations by injecting a high-pressure combination of fluids, chemicals, and proppants causing the fossil fuel to flow to the production well. Proppants are proprietary chemicals that are mixed with sand or gravel used to prop open the fissures created in the rock to allow the gas to flow back to the surface and be extracted from the well.

During the fracking process, millions of gallons of fracking fluid—a mixture of water, sand, and chemicals—are injected into the ground to break up the shale and release natural gas. While each company's formula is a closely guarded secret, in some cases the mix includes known carcinogens such as benzene. Some of the fracking fluid remains underground where it could potentially contaminate groundwater in the future, but much of it is brought back to the surface as wastewater. That wastewater contains fracking chemicals as well as naturally occurring radioactive materials and metals found in the surrounding soil. The wastewater is often pumped into holding ponds where it can leak and settle into surrounding groundwater and impact wildlife. The contamination of groundwater is of major concern for those who live near drilling operations and rely on drinking water wells. And the contamination of watersheds that provide drinking water for millions of people in cities hundreds of miles away from any natural gas drills poses a significant threat as well.

The Marcellus Shale formation, located in the Northeast United States, is of particular interest to the oil and gas industry, not just because of its large, untapped reserve, but also because of its proximity to major population centers. This proximity makes the gas easy to deliver to processing center and aids in distribution of the resource. It is also one of the largest volume formations containing shale gas within the United States (US Energy Information Administration 2016). There are other formations, such as the Eagle Ford formation in Texas that also have large areas of untapped production. Figure 4.10 shows the major shale gas plays in the United States.

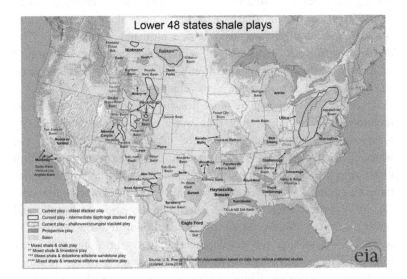

Figure 4.10. Major shale gas plays in the lower 48 United States.

Source: US Energy Information Administration, 2016.

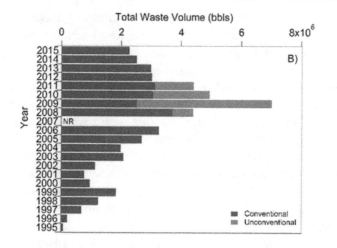

Figure 4.11. Total volume (barrels) of oil and gas wastewater discharged to surface water bodies in Pennsylvania.

Source: Warner et al., 2013.

That proximity to populated areas, however, also raises significant public health concerns. Of primary concern is the potentially damaging impact of natural gas drilling on water resources.

Assessing the impact of natural resource extraction on surface water quality in areas expanding the use of hydraulic fracturing and horizontal drilling technologies is vital for safe and sustainable development (Vidic et al. 2013). The expanded use of fracturing technology in Pennsylvania has led to increased volume of treated water ultimately discharged to rivers and streams. Since 2011, the majority of discharges have been conventional oil and gas wastewater, while unconventional waters have been reused on site or shipped to deep injection wells (Figure 4.11).

Sustainable development strategies for water resources focus on monitoring water temperature and water quality using conductivity measurements, which relate to the total dissolved solids (TDS) in the water. TDS can be derived from natural processes such as migration from underlying sources, dissolution of evaporates, or weathering of aquifer minerals or anthropogenic sources such as oil and gas wastewater, coal flue gas desulfurization discharge, coal mine discharge, and septic and sewer discharges (Warner et al. 2013). Identifying the sources of TDS is vital for assessing the development strategy for water and other natural resources (e.g., hydrocarbons) under changing energy resource development scenarios. Water quality in many regions has become a limiting factor for long-term sustainability and management of water resources (Vengosh 2013).

Elevated TDS concentrations (including bromide) can have important impacts to drinking water treatment facilities downstream that typically received water with TDS much less than 1,000 ppm (Wilson et al. 2012). This wastewater has not only important implications for the energy future of the United States, but also for water quality of communities in areas of resource development.

Water related to oil and gas extraction experiences degradation associated with the discharge of oil and gas wastewater from brine treatment facilities. The TDS of this water can vary from 5,000 ppm to >200,000 ppm with high bromide, chloride, and naturally occurring radioactive material (NORM) (Rowan et al. 2011). There are options for disposing of the high-TDS waste fluids including injection into underground disposal wells, transport to water plants for treatment, and discharge to a nearby surface water body, or reuse for future fracturing. Most commonly, waters are treated and discharged, hence the importance of monitoring devices. These streams also serve as the major drinking water supply for millions of people. With elevated salinity, chloride, and bromide concentrations, the facilities produce greater concentrations of disinfection by-products (Wilson et al. 2013).

Oil and gas extraction is covered under EPA rule 40 CFR Part 435. These rules cover effluent from wastewater discharges from field exploration, drilling, production, well treatment, and well completion activities for oil and gas extraction. Extraction activities are regulated on land, in coastal areas, and in offshore. These rules were last updated in June 2016 to prohibit the discharge of oil and gas extraction wastewater to publicly owned treatment works without pretreatment (EPA 2016).

CHAPTER 5

RISK PERCEPTION AND MANAGEMENT

5.1 INTRODUCTION

Risk is the probability that an event might occur after exposure. All activities comprise some risk; it cannot be eliminated, only mitigated. Risk is identified by first formulating or scoping the problem. If there is a problem with the current condition, then the avenues for altering that condition to a nonadverse effect are explored. At this stage, risk assessment takes place to evaluate possible risk management options.

Risk has several definitions. It can be the probability that an event will occur and project objectives will not be met (project risk), the probability that a performance objective will not be met (performance risk), or the simple consequence of not meeting a goal. A source of risk is any factor that can affect project performance, and risk arises when this affect is both significant and uncertain in its effect on project performance. Project criteria, such as time or project performance goals, have a direct effect on risk. The tighter the time frame or the more stringent performance criteria, the greater the inherent risk of the project. Conversely, projects with a slack time frame or forgiving performance criteria have lower inherent risk. However, project criteria which do not adequately reflect the project goals have a higher degree of risk as they do not acknowledge the need for a minimum level of performance. Risk, at its heart, is the measure of the uncertainty of a project, or a project goal. It can be measured as a probability. People tend to think about probability as a measure of risk every day when reading the weather report—there is a 15% chance or 15% risk that it will rain this afternoon, for example. Operational or project risk differs from the weather example in that it also includes the consequences of failure. This includes any damages that may arise from failure. Therefore, a project may have a low probability of failure, but carry more risk if the consequences of failure are high. Assessing the degree of risk that a project incurs can be difficult, as the damages from failure can only be estimated.

Risk has three primary components: an event, the probability of that event occurring, and the impact of that event. It can be framed that the degree of the uncertainty and damage of the event are a function of the risk incurred. This means that as the uncertainty or damage increases, so does the risk. Risk is the measurement of the degree of foreknowledge of events. The more accurate the estimate of the consequences of an action or the failure of an action to occur, the more accurate the assessment of risk. This is a cumulative effect of all factors on a project outcome. Positive future affects are opportunities, while negative future affects are costs.

Another element of risk is the cause of risk. Some factor, or lack of that factor, induces a change in an element of the project, or a hazard. Hazards are sources of risk that can be minimized and to some degree quantified, if there is enough foreknowledge of the hazard. For example, a driver can swerve to miss a garbage can in the road, if they can see far enough ahead. The garbage can is a hazard, or a source of risk that can be damaging if not avoided. The risk of driving increases as a result of the garbage can in the road, but the future effect of the garbage can is able to be minimized if the driver sees it in time. This leads to the point that risk is also a function of hazards and safeguards. Hazards increase risk, while safeguards decrease it. By identifying hazards and constructing or finding safeguards, risk can be minimized.

There are both external and internal sources of risk. External sources of risk are risks beyond what is managed by the project. Examples of external sources of risk are changes in climate, politics, or regulatory control. This includes legislative safety requirements for consumer or environmental protection. These requirements must be met in the conduct of an enterprise or fines or other consequences will occur. For example, independent testing of Maggi noodles in India revealed high levels of lead in the seasoning. This resulted in fines for noncompliance of the parent company, Nestle India, a 10% fall in the reduction of the stock of the company, and a boycott by consumers. The failure to meet regulatory control added a degree of external risk in the fines for noncompliance and consumer boycott. Internal sources of risk are risks that are in the control of the project management team. This source of risk includes risk arising from project design and established human behavior. Harmful sources of risk are communication or technology failures, project design that does not meet the objective, and poor leadership. Positive sources of internal risk are the performance, skills, and motivation of the project design team. For example, a poorly managed team has a higher degree of internal risk that the project will not be completed on time, while a well-managed team has a lower degree of risk. Internal risk can be managed by risk mitigation and risk contingency.

Risk mitigation is the ability to take actions to minimize risk, such as putting together an appropriate team for the project, while risk contingency is building up reserves or buffering in order to reduce risk. Risk contingency would be having a secure computer backup for project notes in the event of a power failure. There are four categories associated with internal risk: safety, technical, management, and business risk. Each of these risk categories may affect the overall risk of the project. Safety risk is the risk associated with compliance of given standards for the project, such as environmental regulations. This is a manageable risk as the hazard can be mitigated, resulting in a lower risk. Technical risk is the risk that technology or design of the project will not fulfill objectives and can also be minimized. Management risk is the risk associated with human behavior and can add an element of increased risk to the overall measurement. Business risk is the economic opportunity or harm associated with a project. These risks can also impact each other. For example, a technical risk incurred by using a new material can generate business, safety, and management risks. If the new material fails, this results in increased management risk as the team works harder and under a smaller time frame to find a replacement material. If the material does not meet operational or regulatory standards, this results in a safety risk. Finally, a business risk is incurred if the new material does not perform, resulting in economic loss. These risks must all be managed in order to provide the best probability that the project will succeed and therefore the lowest overall risk.

5.2 RISK PERCEPTION

Virtually every aspect of life incurs some risk; how well humans deal with risk depends on how well they understand it. Risk perception is evaluating the severity of the risk and what technical assessment is necessary to characterize it. This encompasses both the risks of existing conditions and the risks of proposed options. It is also the stage at which the acceptable level of uncertainty and variability in assessment is acceptable.

5.3 RISK ASSESSMENT

Risk assessment is the process of identifying the hazard, assessing the exposure, and characterizing the future risk. Risk assessment encompasses the analysis of available data on the hazard in question along with the implications of that data for human risk. For hazard identification, it is sought to determine the adverse health or environmental effects that are associated with the agents of concern. Hazard identification is qualitative data based on available studies of humans and laboratory animals, along with available ancillary data. Examples of ancillary data include structure–activity analysis, pharmacokinetics, and genotoxicity. This descriptor addresses the strength and reliability of current research, mechanistic, and experiential information gained from human and animal study.

Additional assessment of risk is from both dose-response assessment and exposure assessment. For dose-response assessment, evaluation is of degree of adverse effect related to the dose and probability of the occurrence of adverse effects related to a range of doses. Dose-response assessment examines quantitative data between exposure and effects to determine areas of concern. Special attention is paid to the range of potency in animals and humans during dose-response studies as well as correspondence between routes of exposure. Exposure assessment is the evaluation of the exposure or dose of each population under existing conditions and how risk management options may affect the exposure or dose each population receives. The exposure assessment examines the factors pertaining to real-world factors and scenarios of humans who may be exposed to the agent under study. An effective exposure assessment gives the basis for values used in the study and includes data about any assumptions made during assessment. The exposure assessment seeks to illuminate critical areas of uncertainty in the risk assessment, particularly those areas that may pertain to greatest potential contaminant uptake.

Risk identification is an integral part of risk assessment. This involves the integration of qualitative risk analysis techniques. Qualitative risk analysis is a method of understanding risk that cannot be associated with a numeric value, or quantified. This process has three components: objective observation, referencing, and interviewing. Objective observation is the close examination of a current project or objective to help identify risks involved in a new project. This type of analysis relies on understanding of the past project objectives and challenges in order to minimize the risks incurred for a new avenue. Referencing is the consulting of documentation of existing projects and database searching. Referencing helps to provide a comprehensive picture of associated risks from previous work. Interviewing, which may or may not occur, relies on direct consultation with those involved in previous projects and documenting their experiences. Each of these elements makes up the analysis of qualitative risk.

Qualitative risk can be weighted on a simple scale, using the ratings of high, moderate, and low to reflect the degree of risk. High risk means that the risk is likely to impact the schedule, cost, quality, or performance outcomes even with additional support. Moderate risk will cause some disruption, but potential problems may be overcome. Low risks will cause little disruption to the project outcomes, or can be overcome with normal efforts. Other methodologies, such as RISKMAN, use categories; i.e., Category 1, in place of the high, moderate, and low definitions, but the systems are comparable.

Risk quantification is the next part of risk assessment. Risk quantification is a simple mathematic approach to understanding risk. Each risk is assigned two values on a scale. The first value reflects the likelihood of the occurrence of the risk, and the second is the impact of the risk should it occur. These two values (likelihood and impact) are then multiplied to weight the risk. The weight of the risk is often called the hazard, or the hazard value. The larger the hazard value, the greater the risk. This information can then be complied to generate a risk register of all the risk of a project. The register can then be used to discuss the impact of a risk, its nature and areas of prevention for that risk, and any other areas of impact. Other useful columns can be added to the risk register, such as who has ownership of the risk or any secondary effects of the risk. It is specifically noted that risk analysis is an iterative process. Risk should be continually assessed throughout the project. Risk analysis is a subjective measurement, so the same individual should do the risk assessment throughout in order to lower the risk of introducing additional bias, particularly if a project relies heavily on qualitative risk measurements. Use of heuristics and reference to previous projects are good monitoring tools to help assess risk while reducing subjectivity.

Risk mitigation is the ultimate purpose of risk management. Appropriate actions are the paths to success that reduce risk by revising time tables, budget, and other project components in order to reduce and overcome risk. Risk mitigation is also having a comprehensive plan, or risk response in order to meet risks when they occur. Good risk mitigation involves regular assessment of risk and prompt activation of mitigation in order to accomplish project outcomes.

Risk mitigation has two action paths: active and passive. Active risk mitigation is taking steps to reduce the risk or eliminate it entirely before the risk occurs. Active risk mitigation frequently has an upfront cost. For instance, if a specific tool is required for a job, active risk mitigation would be to have a second one of those tools in case there is a problem with the first. Purchasing this second back up tool has an upfront cost associated with it. Passive risk mitigation is waiting until risk occurs. In the previous instance, this would be replacing or repairing the tool if it were to break.

Deciding which type of risk mitigation is appropriate involves a cost–benefit analysis. If it is extremely costly to delay the project should another tool need to be obtained, active risk mitigation would be the best path for this example. However, if the upfront cost of a second tool is greater than the cost of the delay, then passive mitigation would be the better path. It is important to weigh each impact of a project risk in order to make the best decision about risk mitigation. Appropriate control of risk involves calculating or estimating the impact of risk while also considering mitigation strategy and contingency.

Contingency can be thought of as the antidote to risk. A risk management plan that has a lot of contingency will bear very little risk, as each possible risk has an antidote. For example, a project may experience a delay if a manager is unavailable to approve the next step. Good risk management will account for this delay within the schedule, however small it may be. Contingency is most needed in areas of scheduling, materials, and work done by external sources. By planning adequately for these delays, the risk to the project is minimized.

Project risks also depend on the degree of risk the individual in charge is willing to accept. This can be sorted into three categories for tolerance of risk: the risk avoider, the neutral risk taker, and the risk seeker. This boils down to the cost of risk. The risk avoider will demand a premium to accept risk, while the risk seeker will be willing to pay a penalty to accept risk. The risk avoider prefers a certain outcome, while a risk seeker prefers the more uncertain outcome. Attitude toward risk is also important to consider, as risk analysis is subjective.

These two attitudes can be thought of as being governed by specific rules which can be used to quantitatively measure risk. The pessimistic rule, or the conservative decision maker, seeks to minimize the maximum loss or maximize the minimum gain (maximin=minimax). The optimistic rule, or the risk seekers attitude, is to maximize the maximum gain (maximax). Compromise, or risk neutral individuals, fall under the Hurwitz rule: Max (α min + $(1-\alpha)$ max), $0 \leq \alpha \leq 1$, where $\alpha = 1$ pessimistic, $\alpha = 0.5$ neutral, and $\alpha = 0$ optimistic. For each, the highest opportunity value, or the minimum probability of loss, is selected, which will also have an effect on the amount of risk assumed by the project.

Risk identifies, measures, mitigates, and controls uncertainty. It is present in all aspects of project management and can cause a project to fail if not properly managed. The risk management process is iterative and continuous during a project. Effectively, risk management can ensure the success of a project and provide a foundation for successive works.

5.3.1 IDENTIFICATION OF POINTS OF EXPOSURE

Risk assessment also entails the identification of points of exposure. Points of exposure can be large or small, depending upon the potential sources in question. Most commonly, points of exposure are related to the pathways for disease in the human body: mucosal membranes, dermis, and ingestion. Identifications of points of exposure for a toxin are critical to assigning the correct personal protective equipment (PPE) that is required to be worn when in a hazardous environment. Points of exposure from an environmental perspective refer to those areas in which hazardous toxins can migrate from the source. Potential pathways for these exposures include land, water, and air. It is again critical to identify the potential points of exposure in order to choose the relevant mitigation efforts (either treatment or technology) to minimize harmful impacts.

5.3.2 IDENTIFICATION OF CONTAMINANTS OF CONCERN

Identification of contaminants of concern is also critical to the outline of a risk assessment. Contaminants may be concerns as they are a potential human health hazard, wildlife hazard, or environmental hazard. As such, each of these exposures has specific mitigation efforts that must be adhered to in order to minimize risk. Contaminants of concern can be identified at the start of the manufacturing or construction process, or may be identified later after side effects become apparent.

5.4 RISK CHARACTERIZATION

Risk characterization is the nature and magnitude of risk associated with existing conditions. This also comprises the risk decreases (benefits) associated with risk management options. Risk characterization also notes whether risks are increased or decreased and what significant uncertainties exist.

Risk characterization is two-fold. First, it must address the qualitative and quantitative features of the risk assessment. Second, it must also identify uncertainties of the assessment as part of the confidence in the assessment. In risk characterization, the major components of risk (hazard identification, dose-response, and exposure assessment) are summarized along with quantitative estimates of risk to provide an integrated view of the evidence. The risk characterization portion of risk assessment identifies assumptions, scientific consensus, and reasonable uncertainties to elucidate assumptions made during conclusions. It also potentially outlines current research that may clarify the uncertainty in the risk estimation. Risk characterization integrates all information in the risk assessment to communicate the overall meaning of and confidence in the hazard, exposure, and risk conclusions. It balances information from other risk-assessors, EPA professionals, and the public for a thorough characterization of the strengths and limitations of the assessment. Issues are identified by acknowledging noteworthy qualitative and quantitative factors that make a difference in the overall assessment of hazard and risk, and hence in the ultimate regulatory decision. A major source of uncertainty in risk characterization is capturing variability in exposure across populations while balancing scientific assumptions to bridge knowledge gaps. As such, a complete characterization is necessary for an effective risk assessment.

5.5 RISK MANAGEMENT

The goal of risk management is to prevent negative outcomes which may result from exposure based on the risk assessment or perception of risk. Risk management is the evaluation of the relative health or environmental benefit from the proposed options, and explanation of how other decision-making factors are affected by the proposed options. For risk management, each decision and its costs, benefits, and communications are weighed. It also involves the evaluation of the risk options and the assessment of the selected solution.

There are multiple definitions of risk management. Risk management can be the formal process of systematically identifying risk factors, assessing their impact, and providing for their outcomes. It is also identifying and controlling events that have a potential for unwanted change. In the project context, it is the art and science of responding to risk factors throughout the life of the project and with regard to the best outcomes of its objectives. Risk management is proactive with regard to future events and implies control of those events. Good risk management practice develops contingency plans for the objectives of a project and is not reactive. This works by reducing the likelihood of an event occurring as well as the magnitude or impact of the event.

A risk management system has multiple objectives: identifying risk, quantifying risk, and calculating the magnitude of risk. An identified risk is a risk that is shared. By acknowledging and accepting the risk, the project team is better prepared should that risk occur. Quantifying risk is done by forecasting the technical risk involved in the project. This is then used to calculate the magnitude of the risk and determine if it is acceptable. In risk management, it is important to adequately understand the project context. By working in concert with established systems of control, budgeting, and organization, a culture of problem identification and resolution can be established at the foundation of a good risk management system. Risk management can be justified on any project, no matter the size. The level of implementation will vary dependent upon the scale of the project and the relationship to previous work, however. Established programs will require less risk management as the future risks are low and well understood, while new projects will require greater planning and have greater uncertainty.

This process of risk management begins by identifying the context of the project and the project objectives. This includes identifying the interested parties of the project, the project outputs, and the scope and boundaries of the project. This also includes interfacing with any other projects that may overlap the defined objectives. The next step of the risk management process is risk identification. This step is fundamental to the process. This occurs first at a broad level, and then more in depth as each risk is identified. Identified risks are then subject to the subsequent steps of risk assessment, risk treatment, risk review, and risk monitoring. At each phase of the project, risks are reviewed and monitored.

The established methodologies of risk management, such as the DoD risk management guide, list four key stages of risk management. These stages are risk identification, risk quantification, risk response, and risk control. The risk identification stage shows any risks that could jeopardize the project or the project outcomes. These risk factors are then assessed and quantified in the risk quantification stage. If the risk is found to be significant, it is referred up to the management structure, if it is found to be insignificant it is referred down to be discussed and minimized. For any remaining risks, contingency or mitigation plans are developed in the risk response stage. The risk control stage follows directly from the risk response stage as containment of identified risks. This includes risk monitoring and mechanisms for execution of the contingency plans, if necessary.

CHAPTER 6

Scheduling for Impact Mitigation

6.1 NESTING/SPAWNING SEASON IMPACT MITIGATION

It is important to mitigate impacts of development during nesting and spawning seasons to lessen effects on inhabiting species. Increasing energy development, such as unconventional methods of oil and gas extraction, has significant potential for disrupting nesting birds. Areas of sagebrush, home to obligate species such as the sagebrush grouse, are particularly at risk from energy development (Kirol et al. 2015). Nesting success is a primary factor in avian population stability, and declines in nesting success have been linked to population declines.

Mitigation practices promoted by U.S. regulatory agencies follow a hierarchy designed to avoid, minimize, and restore biodiversity on-site while considering offset sites to address residual impacts (USFWS 1981). For oil and gas development, on-site mitigation (i.e., minimize impacts) generally involves redesigning operations and infrastructure, or infrastructure placement with a goal to abate impacts to wildlife. There are a variety of strategies being using by energy developers to minimize wildlife impacts. In areas where on-site operation is noncritical, remote monitoring is used as a spawning/nesting mitigation technique. Reduced vehicle traffic in areas that are remotely monitored reduces fatalities for migration and mating species, as well as chicks (Lyons and Anderson 2003). Other advantages of reducing vehicle traffic include the overall habitat destruction mitigation from road building, reduction of invasive species that use the road as a conduit to seek new habitats, and noise reduction.

Another strategy for minimization of nesting area destruction in energy fields is the reduction of reservoir construction. Man-made reservoirs breed mosquitos which can carry vectors such as West Nile virus (Kirol et al. 2005). Increased vectors have been linked to a decline in nesting success from infection, particularly in sagebrush grouse. Reservoir construction also links nonnative predators such as skunks and raccoons that are associated with riparian areas and predate nests. Reducing reservoir areas can increase nesting and breeding success for those species such as the sagebrush grouse. In areas that are mitigated, nesting success is greater than in those areas that are not mitigated; however, it is still less than in unaltered habitats. In a Wyoming study, there is a 5% difference between each category (i.e., nest survival in mitigated development areas was 5% lower than nest survival in unaltered areas but 5% higher than nests in nonmitigated development) (Kirol et al. 2005). In the same study, it was found that reservoir reduction had a significant positive effect on the nesting success and survival for sagebrush grouse species in the energy development area. In sites where produced water is managed by pipeline transport or perennial drainages, there was an increase in nesting successes. Furthermore, nests within 1.3 km of the water's edge in constructed reservoirs were negatively associated with nesting success (Kirol et al. 2005). Nests along the water's edge were theorized to be less successful due to increased predation by novel nest predators like the striped skunk.

Other strategies for mitigation of nesting impacts in areas of unconventional oil and gas development seek to provide alternative nesting areas for affected species. These artificial nesting areas reduce population decline by encouraging avian nesting away from areas of well production activity and reducing mortality. They also encourage breeding success by providing safer areas for fledgling birds and nest attendance by parent birds. Species that benefit from alternative nesting structures are typically large birds that nest on flat surfaces or rock outcroppings. These structures are mimicked by natural gas structures and birds will nest on the natural gas structures. These nests are often destroyed by operators as they are a hazard to the production facility, reducing breeding success for these birds (Neal et al. 2007). In areas where this nesting behavior is common, a mitigation strategy is the construction of alternative nesting areas. Golden eagles, red-tailed hawks, and ferruginous hawks are all large birds of prey that have been shown to benefit from these alternative nesting sites (Neal et al. 2007). Ferruginous hawks also preferentially reuse nests each year, so nests that may be destroyed during the development of a natural gas site have a detrimental effect on the nesting populations. Alternative nesting sites provide recurring areas safe from development that increase nesting success. Artificial platforms and other anthropogenic structures (i.e., gas condensation tanks, abandoned windmill platforms, power poles) increased nest survival and productivity given the density of human disturbance present in the study area.

Anthropogenic nesting substrates appear not to be ecological traps and have potential use in mitigation; however, effectiveness of artificial nest platforms as a management tool should be evaluated experimentally with pretreatment data incorporated into sampling designs. It is a concern that ferruginous hawks may prefer alternative nesting sites over traditionally available habitat due to accessibility of nesting areas as well as increased prey availability, but this has not been conclusively studied (Neal et al. 2007). Although ferruginous hawks nested in some areas with very little vegetative cover, they produced fewer than the average number of young in areas with >20% shrub cover. Efforts used to maintain shrub cover for other shrub-dependent species of conservation concern, such as the greater sage grouse (Connelly et al. 2000), may, therefore, also sustain or increase the productivity of ferruginous hawks.

Land cover is another important avenue of nesting site mitigation that can help maintain breeding success in areas affected by energy development. Shrub cover such as scrub and sagebrush increases hiding areas for birds in developed areas, provides habitat for prey, and yields nesting materials. Winter habitat for these nesting birds is also important and provided locally by shrub cover. Cover also provides barriers against predators by reducing sound, site, and scent from nests, increasing reproductive success. Nests are typically found in areas of greater cover for sagebrush grouse (i.e., grass associated with nest sites and with the stand of vegetation containing the nest was taller and denser than grass at random sites), suggesting that cover is another significant predictor of nesting success (Connelly et al. 2000). Cover also provides nutrition for nesting birds, as it yields habitat for ants, beetles, and other invertebrates important to the avian diet (Connelly et al. 2000).

Coastal areas are particularly vulnerable during spawning and nesting seasons. The breeding habits of many coastal species coincide with tourist seasons, necessitating extra measures of protection for vulnerable groups such as shorebirds and turtles. Changes to beachfront areas such as structures, whether temporary or permanent, may alter breeding habits and negatively impact species. Negative effects include beach erosion and sand accretion, which may in turn affect crabs or nesting birds (Brown and McLachlan 2002). Direct trampling by tourists can destroy nests, and nesting habitats of turtles or birds promote soil compaction or impact sensitive dune vegetation that helps to maintain beach structure. Management strategies such as zoning, building restrictions, or limits to the number of tourists visiting beaches can mitigate these impacts and reduce negative effects. Other mitigation measures include construction of boardwalks to minimize trampling, building nesting boxes for impacted species such as plovers, and even closing beaches or portions of beaches to reduce impacts.

Disturbances in nesting sites can also be mitigated by construction of alternative nesting areas. Alternative nesting areas can take the form of boxes, fenced areas restricted to human approach, and platforms above the tidal zones. Platforms are the most common form of mitigation for alternate nesting sites as they are simple to build and can even be co-located with original nesting areas. As part of providing alternative nesting areas for mitigation, monitoring programs are also used to ensure that the mitigation is effective and the nesting sites are in use. Volunteers staff the programs which count birds via observation and banding. Bird banding is a simple and successful way to count birds as it reduces the probability that an individual is counted twice. There is frequent difficulty in the assessment of alternative monitoring sites with shorebirds as birds migrate from site to site during the season, as well as hiding chicks when a perceived threat (such as a human observer) is near. Therefore, in order to monitor effectively, volunteers must band birds extensively to reduce dual counting and find observation sites that allow for viewing birds without being perceived. This is important as productivity measurement is integral to determining the success of the implementation of alternative sites as mitigation measures. Mitigation measures like alternative nesting must also be maintained in dynamic areas such as coastal regions. Changing conditions such as beach erosion, tidal events, and storms all take a toll on constructed mitigation measures so maintenance must be provided for as part of the site plan.

On inland waterways, fish spawning is an area that is frequently impacted by human activity and must be mitigated. Dams and other waterway construction in the Pacific Northwest of the United States has reduced or eliminated available spawning grounds for fish species within these areas. These fish species, such as salmon, catfish, and murrel, utilize gravel beds for spawning which have been eliminated by dam construction. Mitigation efforts for spawning to support these fish populations take several shapes, depending upon the natural geography of the area. In areas where it is feasible, spawning grounds are supplemented by gravel augmentation, in which spawning-sized gravel is added to the beds to replace eroded or removed rock (Figure 6.1) (FAO 2012).

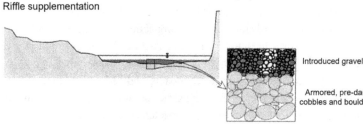

Riffle supplementation

Introduced gravel

Armored, pre-dam cobbles and boulders

Figure 6.1. Schematic of gravel augmentation below a damn within a riffle. *Source:* Bunte, 2004.

To increase the effectiveness of gravel deposition, large boulders and rocks are also added to create pools and riffles for spawning. Manipulating these rocks together creates cover for fish, alters the downstream flow pattern to produce spawning pools, and manages sediment within the stream (FAO 2012). In other areas, spawning can be assisted by the placement of fish passes and fish ladders. Fish passes take the form of rocky passes, pools, or artificial river bypass channels and slots that allow fish to move from lower pools past the dam (FAO 2012). Fish ladders perform that same service by allowing fish to migrate past a blocking dam to return to home grounds to spawn.

In areas where it is not feasible to augment riffles with gravel, or fish passes are insufficient, there are also fish hatcheries that support and assist spawning populations. This is most common with salmon and steelhead fish. These fish are trapped below the dam and transported to the hatchery to spawn. Fish are reared within the hatchery and returned to native streams and pools for sport fishing, tribal lands fishing, and commercial operations. Modern hatcheries mimic naturally occurring ponds by using s-shaped concrete ponds painted with underwater camouflage. These operations, such as that of the Nez Pierce tribe in Idaho, have shown to increase success among released fishes by allowing young to develop with proper markings and to acclimate to their native streams faster than farm-raised fish (Fifer 2005). The Nez Pierce facility uses diverted creek water, submerged logs, and a native insect population to feed and house their fish, which also contributes to the success of their mitigation efforts. The water flow speed, temperature, and volume are altered to mimic natural conditions, better acclimating the fish to their final habitat (Fifer 2005).

6.2 HABITAT DESTRUCTION MITIGATION

Habitat mitigation is activity that compensates for or offsets adverse impacts to habitat in areas of land or energy development. Mitigation preserves, enhances, restores, or creates habitat for flora and fauna affected adversely by development. Mitigation measures are meant to minimize or diminish the effects of development. It is an integral part of reducing project impacts and has no meaning without the project (CEEWeb 2012). Examples of mitigation include animal passages, sound barriers, and reducing functions within the project (IE road speed limits in an area heavily tracked by animals). Mitigation is different from compensation, in that mitigation is not intended to be separate from the project, nor can compensation be the reason the project is authorized (CEEWeb 2012).

In the United States, new development must include a habitat mitigation plan under rules established by the Fish and Wildlife Service in 1981 (USFWS 1981). These rules define habitat in terms of four resource categories, ranging from irreplaceable habitat to habitat of medium to low value for evaluation species (USFWS 1981). The Fish and Wildlife Services reviews project impacts to minimize or eliminate habitat impacts by evaluation of the long-term biological impact of the project, including cumulative effects. The Fish and Wildlife Service recommends mitigation means and measures which are used to guide the actions of developers.

These measures fall into five categories, with recommendations for each plan of action. The first category is to avoid the impact. It is most highly recommended to avoid impacting the site by designing the project to avoid altogether damage or loss of resources by creation of avoidance structures, multiple outlets, and water pollution control facilities (USFWS 1981). Where avoidance is not possible, the next priority recommendation is to minimize the impact. This is the more common pathway recommendation by the Fish and Wildlife Service, and suggests developers locate their project at the least environmentally damaging site, reduce the project size, and schedule timing of the project to minimize disruption of the biologically sensitive community (i.e., avoiding construction during migration or nesting periods). Other recommended avenues of impact minimization include selective tree clearing, timing, and controlling water flow diversions and releases and controlling public access (USFWS 1981). The third priority recommendation is to rectify the impact. Rectifying impact is a mitigation measure that takes place after the project is complete. Actions for rectifying impacts include regrading disturbed areas, reseeding areas where earth has been replaced, planting trees and shrubs to speed vegetative recovery, and to restock fish and wildlife resources (USFWS 1981). Developers are also subject to the fourth priority recommendation of reducing or eliminating the impact over time. In this mitigation action, periodic monitoring is provided to track habitat recovery. Personnel are tasked with monitoring ecology, flora, and fauna to preserve existing or recovered fish and wildlife, maintaining structures such as nesting boxes and fish ladders, and gathering data regarding species recovery. The lowest priority recommendation is to compensate for impacts. Developers compensate for impacts by increasing habitat values of existing areas such as improving nearby public lands (USFWS 1981). They may also provide buffer zones, acquire wildlife easements, and lease habitats to compensate and mitigate for existing losses resulting from project development.

Habitat mitigation is also approached by something called mitigation banking. Mitigation is the preservation, enhancement, conservation, or restoration of a habitat which compensates for adverse impacts to other nearby ecosystems. An example of mitigation banking would be if a corporation constructs a new building on pristine land, impacting habitat, but compensates for that loss by offering the addition of remaining acreage for a park or preserve. The goal of mitigation banks is to replace or offset the value of habitat lost by the project by an equal or greater measure of protected habitat.

In the United States, mitigation banking is governed by the Clean Water Act. Mitigation banks, under the administration of the appropriate agency (U.S. Fish and Wildlife Service, U.S. Army Corp of Engineers, Environmental Protection Agency, etc.), administer the credits and sell them to developers whose projects impact those ecosystems. Credits are units of exchange defined by the ecological value of converting the ecosystem for economic purposes. Credits can be used to compensate for destruction of wetland, wildlife, or riparian habitat. They place a perpetual conservation easement on the land along with a trust fund dedicated to maintenance and preservation of the mitigation-banked area. The credits for mitigation banking must be purchased prior to development (often as part of the permitting process), which ensures that the mitigation request is fulfilled. This often speeds up the permitting process and benefits both the developer and the ecosystem as habitat loss and maintenance of mitigation areas are initially provided, rather than remediated following the development. Mitigation banks help to consolidate fragmented conservation areas into contiguous preserves, which benefit the habitat as well.

When land is cleared for development, the loss of trees must also be mitigated to maintain habitat and prevent further destruction. Mitigation tactics for tree loss fall into two categories: protection of existing trees and planting new trees (and general forest restoration) (Phytosphere). Relative to the area of tree loss, mitigation can take place on-site or off-site (within reasonable proximity to the area of loss). Protection of existing trees on-site as mitigation can happen by relocating structures during project design and siting, utilizing specialized construction techniques to minimize damage to tree roots, and setting aside areas of the project as forest preserves (Phytosphere). Off-site mitigation measures that protect existing forest include purchasing land with existing trees to set aside as a public preserve or land trust administered by a public agency, and establishing conservation easements on tree stands to prevent and protect trees from removal (Phytosphere). Planting new trees on-site as mitigation measures for tree loss puts new trees on landscaped areas and within the on-site area designated as forest preserve. Off-site, new trees are planted to reforest or rehabilitate degraded natural forest, added to existing public lands and trusts, and afforested on former woodland and forest sites (Phytosphere).

There are seven management objectives associated with tree loss that can reduce habitat degradation through mitigation. Preventing net loss of forest canopy or tree type is accomplished by afforestation with an appropriate mitigation ratio. If the ratio of new trees to trees removed is one to one, then there is no net loss of forest type or canopy. Replacement trees must have similar longevity and success rate in order for the mitigation to be successful. Maintenance of the mature tree canopy is another mitigation measure for tree loss. Mature canopy can be maintained by protecting existing trees; a canopy from new trees will take time to mature. Esthetic values of trees as a management strategy can only be mitigated by protecting existing trees; new trees have less aesthetic value until they have matured.

Habitat values associated with an existing forest can be maintained and mitigated in several ways. Values can be mitigated by location of mitigation trees, re-siting of infrastructure to avoid habitat degradation, and reducing the level of disturbance where possible (Phytosphere). Species diversity also needs to be maintained or mitigated to ensure successful forest rehabilitation following tree removal. Locally uncommon or rare tree species should be conserved where possible, and new plantings should reflect initial forest diversity prior to loss. Age diversity can also be mitigated by conservation of existing trees and spaced planting of new trees. Mitigation for age diversity is not successful if all trees are planted at the same time, resulting in an even-aged stand (Phytosphere). Genetic tree resources are mitigated in the same manner as age and species diversity.

Location and ownership are important for mitigation of tree loss. In areas that are locally owned, it is simpler to mitigate loss, but resources may be limited due to the size and availability of the site. In off-site areas that are not locally owned, mitigation resources may be greater, but diversity may be more difficult to preserve. Tree resources and effective mitigation efforts within a project must take into account all factors in order to construct and execute the most effective counter-measures for tree and habitat loss within an area.

Habitat mitigation efforts are especially important when development causes fragmentation of habitat, furthering losses. Building of roads, power lines, train lines, and other linear infrastructure causes habitat fragmentation and destruction, both directly and indirectly. Linear infrastructure directly causes habitat destruction by disrupting wildlife corridors, and increasing wildlife mortality. Indirectly, habitat is destroyed by the alternation of habitat microclimates, incursion of weeds and nonnative plants, and disturbance due to vehicle noise, lights, and movement (Goosem 2004). Mitigation efforts focus on reducing or eliminating these effects, while minimizing the overall disturbance of the project.

Vegetation loss that results in habitat loss can be mitigated in several ways. Project construction can be realigned to minimize direct loss of habitat by altering the project siting or proposed route. In areas with critical habitat (such as with a housing threatened species or unique ecosystem), this is the most desired mitigation option. In areas where the project route cannot be altered, options for offsetting or compensating for linear infrastructure clearing can be taken (Goosem 2004). Other options for reducing habitat vegetation loss include revegetation of the area following construction (Figure 6.2).

Revegetation of cleared areas is often a problematic and ineffective mitigation strategy. This occurs because clearing areas for roads and power lines alters the local hydrology and microclimate, which results in greater production of plants that prefer more sunlight and water (Lee 2006). These plants grow preferentially, crowding out native plants, and prefer less fecund conditions. Increased productivity of nonnative and weedy species adjacent to cleared habitat is a commonly report effect of construction of linear infrastructure, and can be a negative effect of mitigation efforts focused on revegetation. Defoliation of plants near roads is also an effect that can be difficult to mitigate as insects seeking edge habitat increase in numbers near the cleared area (Angold 1997).

Direct and indirect loss of habitat caused by linear infrastructure will also affect fauna. Loss of habitat will decrease the viability of populations by reducing the size of the population that can be supported in that area.

Figure 6.2. Revegetation of hillside slope with fern.
Source: My Biology Blog.

Quantifying the amount of habitat lost directly is relatively self-evident, while indirect losses are more difficult to quantify. A study of the Horned Lark in Illinois, USA, demonstrated lower population densities in farm paddocks within 200 meters of country roads with 300–3000 vehicles per day compared with further away (Clark and Karr 1979). A similar effect was evident near Boston, Massachusetts, USA, where the presence and breeding of grassland birds adjacent to a two-lane highway (15,000–30,000 vehicles per day) were reduced for up to 700 meters, and up to 400 meters when 8,000–15,000 vehicles per day were present (Forman et al. 2002).

Fragmentation of habitat due to construction of linear infrastructure is perhaps the single most important destruction effect that requires mitigation. The division of habitat into smaller fragments results in lower population sizes. When roads act as a complete barrier or selective filter to movement, as is the case for many wildlife species, these smaller populations may not be connected to other populations, and hence they are at a higher risk of extinction (Forman et al. 2002). This is because new individuals or plant communities are unable to supplement a declining population or to reestablish a locally extinct population. The effects are referred to as barrier effects. Road barrier effects are based on road width, traffic volume, and natural species behavior. Species of animals most at risk of population fragmentation due to roads and traffic include species that are unwilling to travel across cleared areas (Forman et al. 2002). Birds that favor large blocks of habitat, such as "forest-interior" species, are more likely to be affected than "edge species" that can not only feed in open areas but also occupy forest. Studies of the movements of understory birds in the Amazonian rainforest found that roads with a 10–30-meter-wide canopy opening typically formed a territorial boundary while birds more frequently crossed a road where the canopy remained intact (Develey and Stouffer 2001).

Many species of invertebrates and amphibians may also be at risk of population fragmentation because of low levels of mobility, avoidance of the unsuitable surface of roads, and the relatively high potential for collision with vehicles (Gibbs 1998). Some species of reptile are at risk because they avoid roads, as was demonstrated in a study tracking migration of blue-tongued lizards in the suburbs of Sydney, Australia, where home range boundaries were aligned with roads, and they actively avoided crossing the roads (Koenig et al. 2001). Study has shown that even narrow roads (e.g., 3 meters in width at maximum) appear to inhibit the movement of small mammals, but did not completely eliminate road crossing by these individuals. The response of species is often specific and even species within the same guild (e.g., small terrestrial mammals) have been shown to display different road-crossing abilities (Goosem 2001).

Figure 6.3. Wildlife crossing over a roadway in the Netherlands.
Source: EPA, 2016a.

The most common mitigation strategy to reduce habitat destruction and fragmentation with linear infrastructure is the construction of wildlife crossings (Figure 6.3). Wildlife crossings maintain habitat connectivity by providing a safe path under or above a roadway or maintain a linear vegetation strip around power line pylons to provide forest canopy connectivity. Crossings may be purpose built for wildlife, or may serve dual purposes, such as water drainage or project access.

Wildlife corridors such as land bridges above roads mitigate the reduction of habitat by reducing rates of road kill, maintaining habitat connectivity, maintaining genetic interchange, and allowing for recolonization of territory by disrupted species and population maintenance (Forman et al. 2002). Crossings increase or maintain the population viability in areas of habitat disruption and are generally consider effective mitigation strategies at the level of the individual animal (where the majority of studies have recorded data) (Forman et al. 2002). The majority of wildlife crossing structures increase the permeability of the road by allowing individual animals to move safely across the road. In this sense, wildlife crossing structures are generally successful at reducing the fragmentation effects of roads and other linear infrastructure for the individual, and are successful measures to mitigation habitat destruction effects.

CHAPTER 7

ECOLOGICAL RESOURCE RESTORATION

7.1 HABITAT VALUATION CONCEPTS AND PRACTICES

Habitat valuation is the process by which policy and management decisions related to habitat account for human and ecosystem needs. By placing a monetary value on ecosystem resources, policy makers can better account for changes that affect human well-being. Ecosystem resources or services are the contributions that a habitat and biological community provide to human lives (NOAA Habitat Conservation). The services of an ecosystem or habitat are directly related to human values. Examples of ecosystem services with direct human benefits are food, medicine, recreation, and storm protection (NOAA Habitat Conservation). Other values are less tangible, such as the ability of a habitat to absorb carbon from the atmosphere and reduce climate change.

Ecosystem services can be sorted into four categories: supporting services, provisioning services, regulating services, and cultural services. Supporting services are the materials necessary for the production of other ecosystem services (NOAA Habitat Conservation). For example, supporting ecosystem services are soil formation and nutrient cycling, as these support healthy flora and fauna. Provisioning services are the products gained from ecosystems, such as food and water. Regulating services are the benefits obtained from processes such as air and climate regulation like water purification from wetlands (NOAA Habitat Conservation). Cultural services are the nonmaterial benefits derived by humans. An example of cultural services is the esthetic values derived from nature (NOAA Habitat Conservation). Table 7.1 lists examples of other ecosystem services important for habitat valuation.

Table 7.1 Ecosystem services important for habitat valuation (NOAA Habitat Conservation)

Indirect benefits to humans		Direct benefits to humans	
Supporting	**Regulating**	**Provisioning**	**Cultural**
Primary production	Gas regulation	Food	Esthetic
Nutrient cycling	Climate regulation	Fresh water	Recreational
Soil formation	Disturbance regulation	Raw materials	Spiritual
Hydrological cycle	Biological regulation	Genetic resources	Historic
Habitat formation	Water regulation	Medicinal resources	Scientific
Pollination	Waste regulation	Ornamental resources	Educational
Seed dispersal	Nutrient regulation		
	Soli retention		
	Disease regulation		
	Flood regulation		
	Water purification		

These ecosystem services can be sorted into two categories: those that directly benefit humans and those that have indirect human benefits. Those that have indirect benefits (the supporting and regulating categories) are more difficult to assign to valuation, as they provide nontangible benefits to humans. Provisioning and cultural services are simpler to value as they have direct human benefits. In ecosystem valuation, direct and indirect human benefits can be reclassified as market and nonmarket values. Those services with market values, such as food like fish from a lake, can be valued easily, while a walk on the beach is an item with a nonmarket value, which is more difficult to quantify economically. Classifying ecosystem services is a way to estimate monetary value when making habitat valuations.

The habitat or ecosystem services valuation process generally has four steps: identifying services that matter, identifying metrics for those services, quantifying service flows, and valuing service flows (NOAA Habitat Conservation). Each of these steps helps to quantify or assign value to the habitat for economic analysis. Depending on the project, different impact

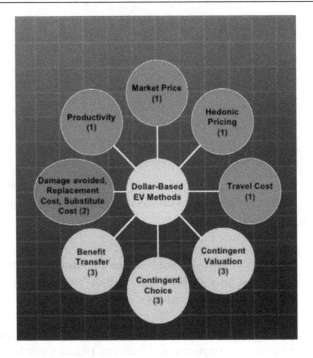

Figure 7.1. Dollar-based ecosystem valuation methods and categories: 1 – market price method, 2 – circumstantial evidence method, and 3 – survey method.

Source: Puma, 2011.

assessment or pricing methods may be used. There are eight different dollar-based habitat valuation methods, which fall into three categories (Figure 7.1). The three categories are market price methods (revealing willingness to pay), circumstantial evidence (implied willingness to pay), and surveys (expressed willingness to pay).

The market price method is the most commonly used method for valuation of provisioning services. This method estimates ecosystem value for products or services that are bought or sold in commercial markets (Ecosystem valuation). It is used to value either the quality or quantity of a good or service, based on market supply and demand. The market price method measures the economic benefit of goods or services supplied or purchased to find the ecosystem valuation. For example, a fishery that is closed by pollution would use the market price method to determine the losses from the consumers and producers impacted by the fishery closure. This value is then equal to the habitat value or benefit gained from cleaning up the pollution and restoring the habitat value (Ecosystem valuation). All areas of ecosystem services gain for removing the pollution, including those with indirect benefits or nonmarket price values (Figure 7.2).

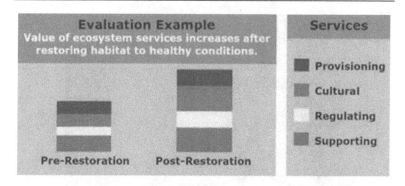

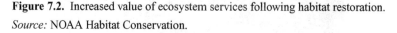

Figure 7.2. Increased value of ecosystem services following habitat restoration. *Source:* NOAA Habitat Conservation.

The market price method, since it is based on data from established markets (such as the price of fish), uses well-defined values. However, data may only be available for a limited number of goods or services in the area, such as a single type of fish. The market price valuation method also only predicts the value for the direct-use resources, and does not estimate the indirect human-use resources well (such as esthetics lost by closure of the fishery shoreline). The market price method also does not account for the other resources necessary to bring the resource to market so it often overestimates the benefits of the resource gained. Estimating the contribution of other resources to bring the item to market can be done using the productivity method, which is another method of ecosystem valuation.

The productivity method, also known as the derived value method, estimates the value of other resources that contribute to the production of a good or service from ecosystem services (Ecosystem valuation). It is frequently used in conjunction with the market price method to better estimate habitat values derived from ecosystem services. The productivity method is a good method for valuation of environmental quality, as it considers the impact of an action on the environment as related to the production of a good or service. For example, a municipal water authority could use the productivity method to estimate the cost of eliminating runoff polluting a water source, as the cost of produced water can be easily determined before and after the removal of the pollutant. The productivity of the water source is affected by pollution and can be assigned a valuation (Ecosystem valuation). Benefits of cleaning up the water source can be directly related to the costs of water purification, which makes the productivity method an ideal choice in this case. The productivity method is also a good way to factor in a natural resource to the cost of production, as changes in its quality or quantity can directly affect the price or quantity supplied of the final good or service (Ecosystem valuation). These changes are then reflected in observable market data which can be used to monetize or value the habitat or ecosystem services utilized by the natural resource. The productivity method may underestimate the value of the natural resources as not all resource benefits are including in the habitat valuation. It is also limited to relationships that directly affect or can be directly related to the production of a good or service. Indirect benefits such as environmental quality benefits from the natural resources are better estimated using the hedonic method of ecosystem valuation.

The hedonic method of ecosystem valuation is the best method for estimating environmental attributes, such as air quality (Ecosystem valuation). The hedonic method is most frequently applied to housing prices. The hedonic method equates the characteristics of a good directly to the services it provides. For example, the hedonic method of valuation for a vehicle would take into account the fuel economy, luxury, comfort, etc. of that vehicle. Since hedonic method is most frequently used for valuation of the environment as related to housing, it accounts for air quality, esthetics, noise, recreation, and other amenities of a site (Ecosystem valuation). The hedonic method is based on available housing data such as pricing, so it is relatively easy to implement. However, the hedonic method relies on understanding the linkage between the pricing of a property and its environmental amenities, which is often not as clear. If the linkage between the property and esthetics or local resource quality is not well understood, the hedonic method underestimates the resource value (Ecosystem valuation). In particular, the hedonic method often undervalues proximity and quality of nearby recreation sites, which necessitates another method of valuation, the travel cost method.

The travel cost method of environmental resource valuation specifically estimates the resource value associated with recreation sites. The travel cost method accounts for recreation site access as well as environmental quality of the recreation site. It uses the time and travel costs incurred by people visiting a recreation site as the cost of access to that site (Ecosystem valuation). The number of trips people make to the site at varying travel costs is the willingness of the group to pay for access to the site and can be understood as a "marketed good" valuation of the site. For example, the justification for the preservation of a recreational fishing site versus developing the site into a hydropower dam would be a good use of the travel cost method of environmental valuation. Comparing the estimated economic value produced by the hydropower dam to the esthetic and recreational value of the fishing area using the travel cost estimation method would determine the environmental valuation of the area. If the recreation and esthetic value exceed the potential economic value of the dam, the dam would not be built. This method is frequently used to protect environmentally sensitive areas from development, such as drilling for oil in national parks. It is a simple method to conduct, as visitors to the recreation site are typically willing to participate in a use survey and data are easy to gain and interpret. The disadvantage of the travel cost method is that it is a general survey method that does not capture greater valuation of the site by visitors who live proximal to it, nor does it well define the relationships between environmental quality and perceived site value. Cost-based methods are better suited for determining the valuations between environmental quality and resource value.

The cost-based methods (damage cost avoided, replacement cost, and substitute cost) estimate the cost of ecosystem services based on the cost of avoiding damages due to lost services, the cost of replacement services, and the cost of providing substitute services (Ecosystem valuation). The cost-based methods do not directly measure ecosystem service value through willingness to pay, like other valuation methods; instead they use the value of replacing services lost as an estimate for the economic value of those services. For example, the cost-based methods may be used to determine the ecosystem value of restoring a wetland for flood protection. Using the damage cost avoided method, the value of the property protected by the wetland or the measures taken to avoid flooding in the absence of the wetland can be used to determine the value of wetland restoration. Similarly, the replacement cost and substitute cost methods would use the value of replacing the wetlands with a retaining wall or the cost of building a levee as a way to estimate the cost of restoring the wetland to provide flood protection by comparison.

All methods assess the environmental services in question as a first step. In the flooding example, this would involve estimating flooding occurrence and the potential property impacts. The damage cost method then calculates the potential damage to property from a flooding occurrence or the amount that people would spend to avoid such damage. The replacement or substitute cost method estimates the least costly means of providing the same service as the ecosystem and whether the public would accept that substitute. In the flooding example, this could be building a levee or retaining wall rather than restoring the wetland to provide flood protection and whether this would be an acceptable solution to the public.

The cost-based methods of valuation have an advantage in that they can provide estimates in areas that are difficult to quantify by providing alternative pathways. As in the flooding example, the difficult to measure value of the wetland can be estimated by the damage cost of replacing flooded property and the cost of building a levee or retaining wall in place of restoring the wetland. However, the cost-based methods do not account for esthetic, recreation, or societal value of the ecosystem resource. They also only represent an aspect or portion of the value of the resource. The contingent valuation method (CVM) is able to include these aspects in its valuation of the resource.

The CVM is the most commonly used environmental or habitat valuation method. The CVM accounts for both use and nonuse values of ecosystem and environmental valuation methods. The CVM directly surveys resource users regarding their preferences for cost of environmental services, or to directly place a value on an ecosystem resource. This method is called the contingent method as it asks people to state their willingness to pay contingent on a hypothetical scenario and description of the environmental service. The contingent method is also referred to as the stated preference method as it relies on the direct responses of those taking the survey, rather than inferring their preferences (Ecosystem valuation). Contingent valuation is most useful for assigning values for nonuse parameters of the environment such as esthetics. It is one of the only ways to place value on these passive resource benefits; however, it is not without issues.

The results of a CVM valuation can be difficult to quantify due to variables in human behavior (such as nonresponse bias), and it can be difficult to implement as it requires academic data-gathering techniques. Also, the results of contingent valuation surveys are highly sensitive to what people are being asked to value as well as the survey context (Ecosystem valuation). Therefore, researchers must clearly define the services and the context and to be able to demonstrate that respondents are stating their values for the services in questions when answering the survey.

It can be costly and nontrivial to design a good contingent valuation survey. A CVM must first choose the appropriate survey sample by determining the extent of the affected population and then choosing the appropriate sample (Ecosystem valuation). The survey must also adequately describe the ecosystem services affected, the specific nature of the change, and the way the change is to be valued in order to gain the best results. For example, a survey concerned with water quality in an area must make it clear to respondents that it is only the water quality in question to be valued, and that a change to the water quality will not affect the overall environmental quality in order to best quantify the value associated with water quality. The survey must also specify the mechanism of payment (monthly addition to bill, increased taxes, etc.) in order to get the best picture of respondent willingness to pay.

The advantage of a contingent valuation study is that it is extremely flexible and can be used with a variety of questions in order to gain data about passive use valuation. A contingent valuation study can take a long time to set up and administer but data gained from the study can be valuable for a variety of functions following the analysis of the study results as it translates passive use values into economic valuations. A CVM has a disadvantage in the many factors and biases that can affect a person's response to the study. For instance, a person who feels strongly about paying taxes may choose a lower amount as a protest response on a valuation that uses increased taxes as a method of willingness to pay which does not affect their true valuation of the resource. Instead, a contingent choice method may be a better method of estimating value for ecosystem resources.

The contingent choice method is similar to the CVM, but it does not ask people to directly state their preferences in dollars. Instead, the contingent choice method asks people to respond to a hypothetical situation by choosing between two groups of environmental services, each with a stated cost. This method uses the tradeoffs that people make when deciding in order to infer a value for the environmental service or characteristic. The contingent choice method is a good valuation method to use when trying to determine the best policy or course of action for a scenario that has multiple sets of actions (Ecosystem valuation).

This method can be used to place value on ecosystem services or it can simply be used as a way to rank decisions in terms of preference or tradeoff. An example would be a mining site in a remote area. As few people visit the site, a travel cost valuation would not be a good way to assign value to the site; instead, a contingent choice method could be used to determine the preferred course of action for the passive or nonuse value of the site. Choices could range from preserving the site as pristine wilderness, to allowing sections of the site to be mined, up to allowing mining for the entire area. People taking the survey would be asked to consider the effects of mining as related to the habitat value of the site, making tradeoffs for each scenario. The tradeoffs made can then be used to assign an economic value to each policy scenario or to simply rank the scenarios to find the most preferred site option.

The contingent choice valuation method can be conducted in a few different ways. The contingent ranking method asks individuals to compare and rank program outcomes and costs in order to rate alternatives in order of preference (Ecosystem valuation). In the discrete choice method, survey respondents are shown two or more alternatives and asked to identify their most preferred choice. In the paired rating method, which is a variation on the discrete choice method, respondents are shown and asked to compare alternatives and rank them in strength of preference. For example, a survey might ask if a respondent prefers diversion of water flow to reduce flooding or creation of a dam to reduce flooding and if that scenario preference is weak, moderate, or strong. Contingent choice may be a better valuation method than contingent valuation as it minimizes some biases by not asking respondents to provide dollar values. By not asking the respondent to make a tradeoff directly between environmental quality and money, but rather asking about tradeoffs, it deemphasizes pricing bias, which allows for the gathering of more meaningful data. Respondents are generally more comfortable ranking attributes or providing qualitative data, increasing the data quality gained in this type of valuation method (Ecosystem valuation). The contingent choice method is an equally complex and difficult method as the CVM, requiring sophisticated study design and data handling, which may make it a nonideal method of ecosystem valuation. Respondents may experience decision fatigue in a large-scale study, or make choices based on those given rather than what may reflect their actual value for the study. The benefit transfer method may be useful in this context as it uses data from actual studies to estimate economic value for ecosystem services.

The benefit transfer method estimates economic value for an ecosystem by using benefits from a similar context. For example, a study on recreational fishing benefits in one state might use the benefit transfer method to estimate ecosystem value by using data from another state to estimate the benefits of recreational fishing in that area. This method is most often used when it is too expensive to conduct a new study or there is too little time available for study, yet data are needed. The benefit transfer method works well for general case studies where value is easy to transfer. For example, researchers studying the benefits of a beach located on a lakefront could use a similar lakefront study to transfer the recreational benefit data. It is important when using the benefit transfer method that the sites are of similar quality and value in order for the data to transfer well. Researchers must also determine that the population demographics are similar for the transferred sample, as well as any adjustments that might need to be made to make the data transferable to the current study. The benefit transfer method is most useful when the quality, quantity, and population demographics of the study site and transfer site are very similar, as are the questions and data regarding environmental quality. The benefit transfer method is a quick and easy method of ecosystem valuation that can be used to gain study data quickly. However, the appropriate constraints and studies for transfer must be chosen in order to gain the most benefit from this ecosystem valuation method.

All ecosystem valuation methods ultimately reach the same goal; to yield data about human behaviors and costs to environment should a set of actions be taken. Each method differs in scope and application, so it is important to choose the method that is most appropriate to the question to be answered. The most common use of ecosystem valuation methods is in benefit–cost analysis. Economists use benefit–cost analysis to determine if the given action will have positive or negative benefits to society as a whole. It is only one of many possible ways to make decisions about the natural environment, and determines only the most economically efficient option. Benefits–cost analysis, coupled with ecosystem valuation, is based on people's preferences or willingness to pay, which may not be either the most societally acceptable option or the most environmentally beneficial one. In order to gain the most complete picture, analysis is supplemented with overriding environmental applications and equity considerations. Specific ecosystems may require special implementation of benefits analysis as well as restoration considerations, both of which are additional considerations when applying benefits analysis and ecosystem valuation.

7.2 WETLANDS RESTORATION AND REPLICATION PROJECTS

Estuaries and wetlands are among the most productive ecosystems in the world. They provide a wide variety of services to people and animals; among them are habitat, flood protection, and water filtration. Unfortunately, up until recently, many of these areas were viewed as wasted land and were drained or filled in for agriculture or similar purposes. This could lead to habitat loss for thousands of shorebirds, water fowl, fish, river otters, and other animals. The effects of this habitat loss are now being felt in the form of greater flooding, loss of biodiversity, and shoreline erosion. Restoration and replication of wetland areas is now a priority for many agencies seeking to mitigate damage to these important natural zones.

Wetland restoration and replication projects seek to restore or replicate areas of wetland that have been destroyed by draining, agriculture, or are otherwise degraded. Hydrology of a wetland is restored by removing drainage lines or tiles, installing dikes to retain water, and building embankments to promote water retention (Minnesota Department of Agriculture 2017) as shown in Figure 7.3.

In other areas, stream channeling, ditching, and disconnection from other shoreline ecosystems may need to be addressed as part of a wetland restoration plan (USEPA 2000). To replicate the natural hydrology of a wetland, it may become necessary to elevate the bottom of the site, or to dredge the area if the wetland has been filled in. Recreating the natural shape of the wetland helps to restore the natural fluxes of nutrients and water cycles. Restoring the wetland base also helps to rebalance the microbiology of the site, which contributes to water filtration and survival

Before and Targeted Post-Hydrologic Restoration Conditions

Figure 7.3. Restoration of natural hydrology by removal of dikes and ditches. *Source:* Griffin et al., 2012.

of native biota (USEPA 2000). These actions help to replicate the wet, swampy ground conditions of a natural wetland.

It is important to provide for passive restoration of the wetland (i.e., reducing or eliminating degradation sources and allowing recovery time), rather than active restoration whenever possible. This course of action helps to increase the likelihood that the restoration of the wetland will persist without human intervention. A self-sufficient restoration plan ensures that the wetland will be able to adapt to future changing conditions such as increased runoff due to flux in impervious upstream surfaces (USEPA 2000). A robust ecosystem will persist if funding sources or other means of human intervention to prevent degradation are eliminated in the future.

It is important to understand the place of the wetland within the greater ecosystem and to understand how other indirect impacts are affecting degradation of the area. For example, upstream development may be causing greater flows to the wetland and increasing degradation. Irreversible changes in the watershed may cause other changes within the wetland that will prohibit restoration to pristine conditions. In this case, it is important to structure the restoration plan to provide the most benefit for the natural potential of the area (USEPA 2000). To assist with restoration efforts, a comparison site may be chosen. This site is ideally a similar pristine wetland area within the same region as the area to be restored. Hydrology may differ, but a comparison site can help to measure site progress and success of the restoration (USEPA 2000).

Restoration of the natural hydrologic regime may be the simplest way to restore a wetland. By providing conditions that are similar or identical to the original soil profile of the area, repopulation of native plants and species may occur without further intervention. Reestablishing stream channel, natural floodplain, and tidal zones allows for regrowth of riparian vegetation and habitat that can be seen as sufficient progress toward restoration (USEPA 2000). This method of wetland regeneration is favored, where possible, as it allows for natural succession and repopulation of the wetland area.

It is important to eliminate nonnative species that may have colonized the degraded area during restoration. Nonnative species can outcompete native plants and animals, inhibiting growth and habitat redevelopment of the wetland area (USEPA 2000). Introduction of nonnative species can occur from disturbance of natural areas during restoration and it is important to minimize their introduction to the most vulnerable areas of the site. This can be done by controlling sources of soil such as hosing off heavy equipment before moving to another site as well as physical removal of invasive species.

Wetland restoration can have other benefits beyond the restoration of natural hydrology along the streambank or shoreline. Functioning wetlands provide shelter for native species, act as natural sinks for pollutants, and provide nutrient cycling to support native plant and animal communities. Bioengineering, which is the method of combining living plants with dead plants or inorganic materials to produce living systems, is a great technique to provide maximum benefit to the restoration project (USEPA 2000). Bioengineering techniques to restore wetland areas have been successful for control of erosion, flooding, and even water treatment. Natural decontamination of runoff and mitigation of storm water are two areas in which bioengineering is a good technique for wetland restoration. Adaptive management plans can help determine whether bioengineering would be a good technique to employ for a specific wetland restoration project (USEPA 2000).

In the United States, the Environmental Protection Agency has many resources and guides to assist with wetland planning and restoration. The Handbook for Developing Watershed Plans to Restore and Protect Our Waters along with the EPA Region 5 Wetlands Supplement: Incorporating Wetlands Into Watershed Planning, are free resources to help communities implement plans for watershed improvement by developing watershed management strategies, including wetland restoration and rehabilitation (USEPA 2012). There are also a number of technical memoranda and data warehouses available free from the USEPA such as the Watershed Tracking, Assessment, and Environmental Results (WATERS) tool that communities may find useful in assisting with their wetland restoration and rehabilitation efforts.

The success of wetland restoration and rehabilitation can be dependent on the type of wetland to be restored. Some types of wetlands are more difficult to restore than others due to the difficulty of reestablishment of woody plants and other flora. Herbaceous wetlands, such as freshwater emergent marshes and wet meadows, have been successful in restoration, while sedge meadows and wet prairies have seen mixed results (National Research Council 2001). This is due to the difficulty in maintaining the proper plant community within the wetland. Freshwater emergent marshes and wet meadows have very diverse and hardy plant communities, while sedge meadows and prairies have more narrow communities of flora. Sedges, while being hardy and woody plants, take time to establish a foothold within an area and can be outcompeted by other species during rehabilitation. Woody plants, in general, require a longer period of establishment and growth. Wetland communities that rely on these plants, such as shrub swamps and forested wetlands, are more difficult to rehabilitate and restore due to the long growth and establishment period for the woody trees and shrubs which form the backbone of the wetland (Figure 7.4) (National Research Council 2001).

It is also difficult to replicate the density of trees and shrubs within the wetland area when replanting the area for rehabilitation. Established woody communities have tall tree heights, low tree density due to competition and mortality, and larger tree widths which allow for lesser sunlight and greater understory growth. It is due to this tree establishment that understory growth and reestablishment in these types of wetlands are often most difficult to replicate. The productive understory within a shrub swamp is a key part of the ecosystem, and the plant community within it is typically water and shady loving. Established forested wetlands can support this community easily, while newly growing trees are not able to provide the necessary shade and protection for the understory.

The wetland that is easiest to rehabilitate are seagrass and salt water marshes. These wetlands are integral parts of coastal ecosystems. They are

Figure 7.4. Restoration of a forested swamp in Lake Long, Louisiana.
Source: RES, 2017.

Figure 7.5. Restoration of salt water marsh in Louisiana.
Source: GeoEngineers, 2014.

regularly flooded by the tides and dominated by salt tolerant herbs, grasses, and shrubs. These wetland areas were among the earliest to be restored and rehabilitated due to their value as nursery grounds, migratory stops for waterfowl, and flood protection (Figure 7.5). In the United States, the Tidal Waters Act as well as the Clean Water Act and the Habitats Directive accord a high level of protection to these ecologically important areas. Salt water marshes share a few characteristics that make their restoration and rehabilitation simpler than other wetland ecosystems.

With regard to the plant community, eelgrass or smooth cordgrass dominates the vegetative growth and both are easily propagated. This single species domination makes establishment of flora simple and cheap to accomplish. Vegetation within salt water and seagrass marshes is buffered by ocean water of relatively constant temperature, salinity, and pH, which lessen environmental variability (National Research Council 2001). These same ocean waters help to disburse animal and microbial communities among distant or fragmented marsh areas. It is also important to note that eelgrass and smooth cordgrass are natural colonizers of bare substrate or "early succession" species (National Research Council 2001). The easy establishment of these grasses helps to create an initial foothold for other colonizing species and continues to support the ecosystem once other plants become established.

It is important to provide sufficient tidal flow to marsh areas that are being restored and rehabilitated, as without sufficient salt content, eelgrass and smooth cordgrass are outcompeted by other plants such as common reed which has a lower tolerance for salt, but favors the same low-nutrient saturated conditions. Reestablishment of tidal hydrodynamics is a critical step in the restoration process as in many areas saltwater and seagrass marshes were filled in or drained. Water and flood control structures also have restricted tidal flows, cutting off the water sources of these marsh areas. When tidal flooding regimes are restored by modifying control structures and breaching dikes and gates, these marsh areas are quick to recover.

Figure 7.6. Vernal pool in Northern California.
Source: US EPA, 2016b.

Other types of wetlands are not so easy to replicate due to complex soil, moisture, and other conditions. Those wetlands that are difficult to rehabilitate or recreate include fens, bogs, and vernal pools. These ecosystems require specific hydrology and climate conditions that make replication difficult. For example, vernal pools are shallow depressions which only hold water for short periods of time (Figure 7.6) (National Research Council 2001).

In the northeastern United States, these pools form during rainy periods in the spring and fall, while in the southwest, they form during winter. Due to their ephemeral nature, it is difficult or impossible to recreate these conditions during rehabilitation. Specific attention must be paid to the soil type and moisture conditions, and sites selected nearest to the original location have shown to be the most successful in recreation. However, colonization of the ephemeral plant and animal community remains difficult. In some areas, workers rehabilitating vernal pools have used vacuums to remove macroinvertebrates and seeds from existing pools to inoculate newly constructed pools, with limited success (National Research Council 2001). Fens, which are productive wetland ecosystems that rely on groundwater seepage and calcium-rich soils, are also difficult to recreate (Figure 7.7).

Groundwater is a resource that grows scarcer within the American west, and drawdowns of the natural aquifer by cities and towns mean that in many places, the water table no longer reaches the surface. Fens rely on surface water seepage from groundwater aquifers to support dense

Figure 7.7. Fen and wet meadow complex restoration between earthwork in 1998 (A), planting in 1999 (B), early plant growth in 2003 (C), and mature sedge, willow, and moss dominated vegetation in 2013 (D).

Source: Chimner et al., 2016.

stands of herbaceous plants that are adapted to oligotrophic conditions (National Research Council 2001). The rock in areas that support fens is typically low in nitrogen and phosphorous and high in calcium, resulting in a high pH water that is difficult to recreate. Surface water runoff, which is typically high in nutrients from agriculture and oxidized sulfur, can irreparably harm the unique soil and water chemistry of a fen, as well as restoration efforts. Bogs are wetland areas that are difficult to restore due to unique water chemistry and vegetation conditions (Figure 7.8).

Typically formed when water is trapped within an area that is high in organic matter, bogs are difficult to recreate due to the slow formation of their ecosystems (National Research Council 2001). Decaying plant materials accumulated over thousands of years make up the substrate of the bog. Therefore, when one is disturbed, it is nearly impossible to recreate. The unique acid conditions of bogs also make their ecosystems difficult to recreate and rehabilitate.

Wetland ecosystems can be restored and rehabilitated given the proper attention to hydrology, plant community, soil conditions, and other unique features. While some types are more difficult than others to restore, it is possible to repair damage done by anthropogenic sources with sufficient investment of time and resources. These productive ecosystems

Figure 7.8. Restored cranberry bog, New Jersey pine barrens.
Source: New Jersey Conservation Foundation, 2017.

provide food and shelter to plants and animals, recreation, esthetics, and engineered benefits, such as water quality treatment, all of which benefit humans. Restoration and rehabilitation of these unique ecosystems will allow future individuals to reap their functions and benefits.

7.3 FOREST RESOURCE RESTORATION AND REPLICATION

Forest restoration and replication is the process of restoring and maintaining ecological function within degraded forest areas. Globally there are over two billion hectares of deforested land and degraded forest area with potential for restoration (IUCN 2016). The United States Forest Service recognizes the interconnected landscapes that forest areas encompass and packages its programs under a single budget line item called the Integrated Resource Restoration (IRR) program, co-administered by the United States Department of Agriculture (USDA 2004). The forest resource restoration program has been administered this way since 2012. This approach allows for multiple programs and phases of ecological restoration to occur under the Forest Service umbrella to rehabilitate and replicate a whole ecosystem, rather than just the forest. The Forest Service recognizes the ecological complexity of a forest and its integration within other surrounding ecosystems. Through the IRR, the Forest Service can undertake restoration and replication of forest on multiple fronts, including restoration of forest understory such as moss and mushrooms and native plant development (USDA 2004). The integrated approach hopes to restore 15 million acres by the year 2020.

Restoration and replication of forest lands can take many forms. New tree plantings, managed natural regeneration, agroforestry, and improved land management, including agriculture, are just a sampling of the activities that can fall under the restoration of forest resources (IUCN 2016). Other concerns of forest restoration include climate change, bark beetle infestation, invasive species, and fire management. Forest restoration and replication within the United States is an integrated program with many partners in order to serve the whole ecosystem. For example, work on using forests for sequestration of carbon dioxide to reduce climate change partners with the USDA, the Forest Service, the Environmental Protection Agency, academic institutions, and the private sector and is administered by the Department of the Interior (IUCN 2016). Goals of the carbon sequestration initiative include long-term capture and storage of carbon dioxide, improved biodiversity and wildlife habitat condition and connectivity, improved water quality and quantity, reduced soil erosion, decreased invasive species, improved environmental esthetics, and enhanced recreational experiences (IUCN 2016).

The Healthy Forest Initiative is another multiagency effort for restoration of forest resources focused on fire management. This effort focuses on the reduction of wood fibers and biomass within forests to reduce available fuel for wildfires. This is due to years of drought and build-up of forest fuels on federal lands resulting in uncharacteristically severe wildfires. Modern fire control, which reduced or eliminated small wildfires along with unhealthy forest growth and inactive management, has exacerbated the problem (USDA 2004). The Healthy Forest Initiative is an active management strategy which includes thinning and prescribed use of fire. The goal of this initiative is to restore ecological health and reduce hazardous fuels build-up for better forest management.

The western bark beetle is another area in which the Forest Service is refining its management strategy. Across the western United States, bark beetle infestation is killing trees at an unprecedented rate (Figure 7.9). These dead and felled trees are then fuel for catastrophic wildfires. While the bark beetle and spruce beetle are natural parts of the forest ecosystem, recent infestations have reach unprecedented numbers (US Forest Service 2015).

In 2009 alone, more than eight million acres were newly infested by the mountain pine bark beetle. Across vast acres in the West, even-aged stands of pine forests have formed as a result of years of fire suppression and large-scale, intense logging at the turn of the century. Many of these tree species life histories are fire-adapted, and lodgepole pine, for

Figure 7.9. Bark beetle infestation in the western United States.
Source: US Forest Service, 2015.

example, naturally regenerates in the presence of fire (US Forest Service 2015). These homogeneous and overly dense forests have provided an extensive food source for beetles, and they have responded with large population build-ups. In addition, climate change has resulted in warmer winters that have not been cold enough to reduce beetle populations. This phenomenon, combined with multiyear drought, has allowed beetles to proliferate at higher elevations and latitudes and has resulted in more beetle generations per year in some areas. On average, yearly bark beetle-caused tree mortality is about equal to wildland fire tree mortality across the United States (US Forest Service 2015).

Trees felled by bark beetles retain their needles within the dead tree crown, increasing the risk of fire. These needles stock the canopy with dry, fine fuels that can ignite quickly during weather conditions conducive to fire. Canopy fires are notably difficult to suppress. The overall risk posed by fire temporarily decreases after the dead needles have fallen while the trees remain standing (0 to 10 years after the trees are attacked). From 10 to 20 years onward, the fire hazard increases again (US Forest Service 2015). As dead branches and trees fall, a heavy fuel bed is created, which poses an increased risk of a surface fire. The outbreak increases the number of acres of municipal watersheds and wilderness areas in need of treatment to protect communities and infrastructure from fire. Additionally, due to the lack of safe egress and intense burning conditions created by standing beetle killed trees or down heavy slash, fighting these types of fires is extremely dangerous to fire fighters.

Tree mortality resulting from infestations have also impaired the ability of forests in high-elevation watersheds to provide shade and shelter that help to maintain the winter snow pack and prevent quick runoff during the spring melt and summer storms (US Forest Service 2015). In some areas, the existence of high-elevation, five-needle pine species are severely threatened by bark beetles, white pine blister rust, and climate change. Management strategies that address the bark beetle problem include thinning stands of trees to lower tree densities, and increasing species and stand diversity to help reduce susceptibility to bark beetles. For example, healthy ponderosa pine forests need to be thinned over time by either wildfire or mechanical harvest to reduce density, and thereby decrease vulnerability to beetle attack (US Forest Service 2015). The health of lodgepole pine forests, whose stands are naturally dense, is improved when there is a variety of stand ages positioned across a landscape.

Climate change has played a significant role in the bark beetle infestation. In many areas of the American west, persistent drought has left trees weak and unhealthy. A healthy tree can usually beat back invading beetles by deploying chemical defenses and flooding them out with sticky resin, but just as dehydration makes humans weaker, heat and drought impede a tree's ability to fight back—less water means less resin. In some areas of the Rocky Mountain West, the mid-2000s was the driest, hottest stretch in 800 years. In some areas, there are trees that are adapting to the beetle infestation. These trees are better adapted to hotter climates and seem to survive infestations better due to their genetic adaptations (US Forest Service 2015).

In other forest areas, the focus of restoration is the elimination or reduction of invasive species. Invasive species are nonnative species of insects, plants, and other invertebrates that negatively affect the health and productivity of a forest. These species can range from pathogens and algae to vertebrates and terrestrial species. Examples of invasive species in the U. S. forests include gypsy moths, emerald ash borers, and white pine blister rust (US Forest Service 2015).

The U.S. Forest Service follows a four-pronged approach to management of invasive species; prevention, detection, control and management, and restoration and rehabilitation. This strategic management framework breaks the problem down in manageable parts to address and eradicate the invasive species. The restoration and rehabilitation portion of the framework is the final piece, and provides a pathway for the forest to recover and grow following the successful eradiation of the invasive species. This portion of the management framework ensures the resilience of the forest for years to come.

Restoration and rehabilitation of the forest includes four sub-actions within the management framework: identify and prioritize restoration needs; take action to restore, monitor, and maintain affected areas; assess effectiveness of rehabilitation and restoration activities; and develop, synthesize, and evaluate effective rehabilitation and restoration tools, technologies, and methods. Each sub-action looks to minimize the negative effects of the invasive species and to help the forest recover to at or beyond previous health and productivity (US Forest Service 2013).

The first sub-action helps to prioritize the most critical needs of the forest for rehabilitation. Examples of these critical needs include critical water resources, wildlife habitat, and areas that are more susceptible to fire and erosion. Prioritizing area for rehabilitation lays out a management plan that strengthens the overall forest and ensures high-quality resources receive the most attention.

Following prioritization, action is taken to restore and rehabilitate areas that are most affected by invasive species. The intervention is meant to provide the most efficient and effective treatment for ecosystem health (US Forest Service 2013). Monitoring of affected and treated areas following the actions helps to increase effectiveness and longevity of treatment. Following treatment, assessment of the effectiveness is taken. This sub-action evaluates the success of the eradication or management strategy and helps guide future actions. Recording and assessing treatment strategies help grow knowledge about effective management strategies that can be transferred to other rehabilitation and restoration projects (US Forest Service 2013). This knowledge is used to inform the final sub-action of development of effective tools, technologies, and methods for best practices.

Each sub-action within the rehabilitation and restoration framework helps to support the overall restoration of the forest resource. Successful implementation ensures productive and healthy forest resources for years to come and helps to inform future management decisions. Restoration and rehabilitation of forests is essential to overall ecosystem health and effective strategies and knowledge building ensure ecosystem integrity and strength for the future.

BIBLIOGRAPHY

American Geosciences Institute. (n.d). *Environmental Risks of Mining*. http://web.mit .edu/12.000/www/m2016/finalwebsite/problems/mining.html, (August 11, 2017).

American Geosciences Institute. (2017). *What are Environmental Regulations on Mining Activities? American Geosciences Institute*. N.p. https://www .americangeosciences.org/critical-issues/faq/what-are-regulations-mining-activities, (August 11, 2017).

Angold, P.G. (1997). "The Impact of a Road Upon Adjacent Heathland Vegetation— Effects On Plant Species Composition." *Journal of Applied Ecology* 34, pp. 409–417.

Aral Sea's Eastern Basin Is Dry for First Time in 600 Years. (October 2, 2014). *Aral Sea's Eastern Basin Is Dry for First Time in 600 Years*. N.p. http://news. nationalgeographic.com/news/2014/10/141001-aral-sea-shrinking-drought-water-environment/#/84282.jpg, (August 11, 2017).

Ausness, R. (1978). "Water Use Permits in a Riparian State: Problems and Proposals." http://uknowledge.uky.edu/cgi/viewcontent.cgi?article=1223&-context=law_facpub, (August 11, 2017).

Australian Government. (March 2014). Aquifer connectivity within the Great Artesian Basin, and the Surat, Bowen and Galilee Basins. Background Review, Office of Water Science. Department of the Environment, Canberra.

Basics of an Open Pit Mine. (2012). N.p. http://www.mine-engineer.com/mining /open_pit.htm, (August 11, 2017).

Brown, A.C., and A. McLachlan. (2002). "Sandy shore ecosystems and the threats facing them: Some predictions for the year 2025." *Environmental Conservation* 29, no. 1, pp. 62–77.

Bunte, M. (2004). *Our Work*. N.p. https://www.fws.gov/leavenworthfisheriescom-plex/OurWork.cfm, (August 11, 2017).

CEEweb. (2012). http://www.ceeweb.org/wp-content/uploads/2012/01/Compen-sation_guidance.pdf, (August 11, 2017).

Chesapeake Bay Foundation. (2016). "Solutions: Agriculture." http://www.cbf .org/issues/polluted-runoff/solutions/agriculture.html, (August 11, 2017).

Chimner, R.A., D.J. Cooper, F.C. Wurster, L. Rochefort. (2016). "An Overview of Peatland Restoration in North America: Where Are We after 25 Years?" *An Overview of Peatland Restoration in North America: Where Are We after 25 Years? (PDF Download Available)*. N.p. https://www.researchgate.net

/publication/308691296_An_overview_of_peatland_restoration_in_North_America_Where_are_we_after_25_years, (August 11, 2017).

Clark, W.D., and J.R. Karr. (1979). "Effects of highways on red-winged blackbird and horned lark populations." *Wilson Bulletin* 91, pp. 143–145.

Connelley, J.W., K.P. Reese, R.A. Fischer, and W.L. Wakkinen. (2000). "Response of sage grouse breeding population to fire in southeastern Idaho." *Wildlife Society Bulletin* 28, pp. 90–96.

Crops and Drops. (2005). "Overuse and Misuse." http://www.fao.org/docrep/005/Y3918E/y3918e05.htm, (August 11, 2017).

Develey, P.F., and P.C. Stouffer. (2001). "Effects of roads on movements by understory birds in mixed- species flocks in central Amazonian Brazil." *Conservation Biology* 15, pp. 1416–1422.

Didham, R.K. (2010). Ecological consequences of habitat fragmentation. *eLS*.

Dittmar, T., R. Jaffé, D. Yan, N. Jutta, V.V. Anssi, S. Aron, G.M.S. Robert, and C. John. (2012). "Global charcoal mobilization from soils via dissolution and riverine transport to the oceans." *Science* 340, no. 6130, pp. 345–347.

The Economist. (2014). "Tropical Forests A Clearing in the Trees." *A Clearing in the Trees*. https://www.economist.com/news/international/21613327-new-ideas-what-speeds-up-deforestation-and-what-slows-it-down-clearing-trees, (August 11, 2017).

Ecosystem Valuation—Dollar-Based Methods. (n.d.). N.p. "Market Price Method Estimates Economic Values for Ecosystem Products or Services That Are Bought and Sold in Commercial Markets." http://www.ecosystemvaluation.org/dollar_based.htm. (August 11, 2017).

Environmental Code of Practice for Metal Mines. (2017). "Canada, Environment Government of, and Climate Change Canada." *Environment and Climate Change Canada—Acts & Regulations—Environmental Code of Practice for Metal Mines*. N.p. https://www.ec.gc.ca/lcpe-cepa/default.asp?lang=En&n=-CBE3CD59-1&offset=5, (August 11, 2017).

Eye On Calderdale—Watercourses and Riparian Ownership. (2015). "Eye On Calderdale Watercourses and Riparian Ownership." http://eyeoncalderdale.com/community/watercourse-and-riparian-ownership, (August 11, 2017).

FAO Fisheries and Aquaculture Information and Statistics Service. (2009). "Aquaculture production: quantities 1950–2005." *FISHSTAT Plus—Universal software for fishery statistical time series*. Rome: FAO. www.fao.org/fi/statist/FISOFT/FISHPLUS.asp.

FAO. (2012). "4. Recommended Mitigation Measures." http://www.fao.org/docrep/005/y3994e/y3994e0j.htm, (August 11, 2017).

Field, J., R. Maier, and A. Robert. (2016). *Pollutants Threaten the Everglades' Future EARTH Magazine*. N.p. https://www.earthmagazine.org/article/pollutants-threaten-everglades-future, (August 11, 2017).

Fifer, B. (2005). "Hatcheries & Mitigation." *Hatcheries & Mitigation Discovering Lewis & Clark ®*. N.p. http://www.lewis-clark.org/article/2303, (August 11, 2017).

Footprints in the Dust. (August 11, 2017). *Footprints in the Dust: Environmental Consequences of Open Pit Mining*. N.p. http://sob-leaningleft.blogspot.com/2012/03/environmental-consequences-of-open-pit.html, (August 11, 2017).

Forman, R.T., B. Reineking, and A.M. Hersperger. (2002). "Road traffic and nearby grassland bird patterns in a suburbanizing landscape." *Environmental Management* 29, pp. 782–800.

GeoEngineers. (2014). "Earth Sciences." *Saving Louisiana's Coastal Wetlands GeoEngineers.* http://www.geoengineers.16penny.net/blog/saving-louisianas-coastal-wetlands, (August 11, 2017).

Gibbs, J.P. (1998). Amphibian movements in response to forest edges, roads, and streambeds in southern New England. *The Journal of Wildlife Management* 62, p. 584.

Global Forest Resources Assessment Report. (2005). *FRA 2005—Global Tables.* N.p. 2005. http://www.fao.org/forestry/32006/en/, (August 11, 2017).

Goosem, M. (2001). Effects of rainforest roads on small mammals: inhibition of crossing movements. *Wildlife Research* 28, pp. 351–364.

Goosem, M. (2004). "Linear Infrastructure in the Tropical Rainforests of Far North Queensland: Mitigating Impacts on Fauna of Roads and Powerline Clearings." In *Conservation of Australia's Forest Fauna.* 2nd ed, ed. D. Lunney. Mosman, NSW: Royal Zoological Society of New South Wales, pp. 418–434.

Griffin, M., M. Richard, and B. Martha. (2012). "Getting the Water Right: Practical Experience in Large-Scale Wetlands Restoration." *Getting the Water Right: Practical Experience in Large-Scale Wetlands Restoration—Access Water.* N.p. http://www.ch2mhillblogs.com/water/2012/06/07/practical-experience-in-large-scale-wetlands-restoration/, (August 11, 2017).

Guerin, K. (2003). "Property Rights and Environmental Policy: A New Zealand Perspective." Heap Leaching. *Heap Leaching—BioMineWiki.* N.p. http://wiki.biomine.skelleftea.se/wiki/index.php/Heap_leaching, (August 11, 2017).

Holl, K., C.E. Zipper, and J. Burger. (2002). "Recovery of Native Plant Communities After Mining." https://pubs.ext.vt.edu/content/dam/pubs_ext_vt_edu/460/460-140/460-140_pdf.pdf, (August 11, 2017).

International Cyanide Management Institute. (2012). *Cyanide Facts.* http://cyanidecode.org/cyanidefacts.php, (June 21, 2012).

IUCN. (2016). "Forest Landscape Restoration." *Forest Landscape Restoration.* N.p. https://www.iucn.org/, (August 11, 2017).

IUPAC. (2010). "History of Pesticide Use." http://agrochemicals.iupac.org/index.php?option=com_sobi2&sobi2Task=sobi2Details&catid=3&sobi2Id=31, (August 11, 2017).

Kideghesho, J.R., J.W. Nyahongo, S.N. Hassan, T.C. Tarimo, and N.E. Mbije. (2006). "Factors and ecological impacts of wildlife habitat destruction in the Serengeti ecosystem in northern Tanzania." *African Journal of Environmental Assessment and Management* 11, pp. 17–32.

Kings River Conservation District. (2008). *KRCD—Agricultural Management Practices.* http://www.krcd.org/water/water_quality/ag_mgt_practices.html, (August 11, 2017).

Kirol, C.P., A.L. Sutphin, L. Bond, M.R. Fuller, and T.L. Maechtle. (2015). "Mitigation effectiveness for improving nesting success of greater sage-grouse influenced by energy development." *Wildlife Biology* 21, no. 2, pp. 98–109. http://doi.org/10.2981/wlb.00002 theme/forests/our-work/forest-landscape-restoration.

Koenig, J., R. Shine, and G. Shea. (2001). "The ecology of an Australian reptile icon: how do blue-tongued lizards (Tiliqua scincoides) survive in suburbia?" *Wildlife Research* 28, p. 215.

Lee, E. (2006). The ecological effects of sealed roads in arid ecosystems. University of New South Wales.

LosApos. (2017). *The World's Deepest, Biggest and Deadliest Open Pit Mines.* http://www.losapos.com/openpitmines.

Lottermoser, B. (2012). *Mine Wastes: Characterization, Treatment and Environmental Impacts.* New York, NY: Springer, p. 400.

Lyon, A.G., and S.H. Anderson. (2003). "Potential gas development impacts on sage-grouse nest initiation and movement." *Wildlife Society Bulletin* 31, pp. 486–491.

Maupin, M., K. Joan, H. Susan, L. John, B. Nancy, and L. Kristin. (2014). "Estimated Use of Water in the United States in 2010." *USGS Circular 1405: Estimated Use of Water in the United States in 2010.* N.p. https://pubs.usgs.gov/circ/1405/, (August 11, 2017).

McLaughlin, A. and P. Mineau, (1995). "The impact of agricultural practices on biodiversity." *Agriculture, Ecosystems & Environment* 55, no. 3, pp. 201–212.

Millennium Ecosystem Assessment (MA). (2005). "Nutrient Management," In *Ecosystems and Human Wellbeing: Policy Responses* (volume 3), eds. K. Chopra, R. Leemans, P. Kumar, and H. Simons. Washington, DC: Island Press, pp. 295–311.

Minnesota Department of Agriculture. (2017). *Wetland Restoration.* http://www.mda.state.mn.us/protecting/conservation/practices/wetlandrest.aspx, (August 11, 2017).

Moore, S.K., V.L. Trainer, N.J. Mantua, M.S. Parker, E.A. Laws, L.C. Backer, and L.E. Fleming. (2008). "Impacts of climate variability and future climate change on harmful algal blooms and human health." *Environmental Health* 7, no. 2, S4.

My Biology Blog. (n.d.). *My Biology Blog: Tropical Rain Forest.* N.p. http://18-winter.blogspot.com/2015/06/tropical-rain-forest.html, (August 11, 2017).

National Geographic. (October 9, 2009). "Deforestation." *Deforestation Facts, Information, and Effects.* N.p. http://www.nationalgeographic.com/environment/global-warming/deforestatio, (August 11, 2017).

National Research Council. (2001). "Compensating for Wetland Losses Under the Clean Water Act (2001)." *2 Outcomes of Wetland Restoration and Creation Compensating for Wetland Losses Under the Clean Water Act The National Academies Press.* https://www.nap.edu/read/10134/chapter/4#27, (August 11, 2017).

Natural Resources Canada. (2010). *Mining Sector Performance Report 1998–2008.* http://www.nrcan.gc.ca/minerals-metals/publications-reports/3398#es, (May 31, 2012).

Neal, M., S. Jeff, and S. Steven. (2007). "Artificial Nest Structures as Mitigation for Natural-Gas Development Impacts to Ferruginous Hawks (Buteo Regalis) in South-Central Wyoming." N.p. https://www.blm.gov/nstc/library/pdf/TN434.pdf, (August 11, 2017).

New Jersey Conservation Foundation. (2017). "Restored Cranberry Bog NJ Pine Barrens, by Emile DeVito, New Jersey Conservation Foundation." *Restored Cranberry Bog NJ Pine Barrens, by Emile DeVito, New Jersey Conservation Foundation—Natural Areas Association.* http://naturalareas.org/journal /restored-cranberry-bog-nj-pine-barrens-by-emile-devito-new-jersey-conserv ation-foundation/, (August 11, 2017).

NOAA Habitat Conservation. (n.d.). *Valuing Nature's Benefits.* N.p. http://www .habitat.noaa.gov/abouthabitat/valuingnaturesbenefits.html, (August 11, 2017).

Noss, R. (2002). *Ecology and Biodiversity: The Ecological Effects of Roads.* http:// www.eco-action.org/dt/roads.html, (August 11, 2017).

NPR Water Wars: Who Controls The Flow? (2013). *Water Wars: Who Controls The Flow?: NPR.* N.p. http://www.npr.org/2013/06/15/192034094/rivers-run-through-controversies-over-who-owns-the-water, (August 11, 2017).

Ortiz, O., F. Castells, and G. Sonnemann. (2009). "Sustainability in the construction industry: A review of recent developments based on LCA." *Construction and Building Materials*, 23, no. 1, pp. 28–39.

PEFC. (2017). "About PEFC." *Overview.* https://www.pefc.org/about-pefc /overview, (August 11, 2017).

Petersen, J., and D.G. Dixon. (2002). "Thermophilic heap leaching of a chalcopyrite concentrate." *Minerals Engineering* 15, no. 11, pp. 777–785.

Phytosphere. (n.d). "Guidelines for Developing and Evaluating Tree Ordinances." *Tree Ordinances—Mitigating for Tree Loss.* http://phytosphere.com/treeord /mitigation.htm, (August 11, 2017).

Plowright, R.K., P. Foley, H.E. Field, A.P. Dobson, J.E. Foley, P. Eby, and P. Daszak. (2011). "Urban habituation, ecological connectivity and epidemic dampening: the emergence of Hendra virus from flying foxes (Pteropus spp.)." *Proceedings of the Royal Society of London B: Biological Sciences* 278, no. 1725, pp. 3703–3712.

Polster, D. (2015). "Effective Strategies for the Reclamation of Large Mines." N.p. http://www.eco.gov.yk.ca/pdf/SCOPeDavePolster.pdf, (August 11, 2017).

Puma, S. (2011). "Presentation: Valuing Ecosystem Services, Methods and Practices." *Presentation: Valuing Ecosystem Services, Methods and Practices.* N.p. Web. https://www.slideshare.net/stevepuma/presentation-valuing-ecosystem-services-methods-and-practices, (August 11, 2017).

Rabalais, N.N., R.E. Turner, R.J. Diaz, and D. Justic. (2008). "Global change and eutrophication of coastal waters." *ICES Journal of Marine of Science* 66, no. 7, pp. 1528–1537.

Rawson, D.M., and J.F. Moltmann. (1995). *Applied ecotoxicology.* CRC Press.

ReliefWeb. (April 11, 2000). "The Baia Mare Gold Mine Cyanide Spill: Causes, Impacts and Liability." *The Baia Mare Gold Mine Cyanide Spill: Causes, Impacts and Liability—Hungary ReliefWeb.* http://reliefweb.int/report /hungary/baia-mare-gold-mine-cyanide-spill-causes-impacts-and-liability, (August 11, 2017).

RES. (2017). "Wetlands." *Solutions.* N.p. https://res.us/solutions/wetlands/, (August 11, 2017).

Righelato, R., and D.V. Spracklen. (2007). "Carbon Mitigation by Biofuels or by Saving and Restoring Forests?" *Carbon Mitigation by Biofuels or by Saving and Restoring Forests? Science.* American Association for the Advancement of Science. http://science.sciencemag.org/content/317/5840/902, (August 11, 2017).

Rowan, E., M. Engle, C. Kirby, and T. Kraemer. (2011). Radium content of oil and gas field produced waters in the Northern Appalachian Basin- Summary and discussion of data. US Geological Survey.

Saenger, P., D. Gartside, and S. Funge-Smith. (2013). A review of mangrove and seagrass ecosystems and their linkage to fisheries and fisheries management.

Sholarin, E.A., and J.L. Awange. (2016). *Environmental Project Management: Principles, Methodology, and Processes.* Springer.

Surface Mining Control and Reclamation Act. (1977). N.p. https://www.osmre.gov/lrg/docs/SMCRA.pdf, (August 11, 2017).

Telmer, K. (2006). *Mercury and Small Scale Gold Mining—Magnitude and Challenges Worldwide, International Conference on Managing the International Supply and Demand of Mercury, in Brussels, Global Mercury Project.* http://www.globalmercuryproject.org/documents/non_country%20specific/Telmer%20Brussels%20ASM%20and%20Mercury%20Magnitude%20and%20Challenges.pdf, (July 31, 2012).

Threats to Estuaries. (2016). http://biome-estuaries.weebly.com/threats-to-estuaries.html, (August 11, 2017).

Trautmann, N., P. Keith and W. Robert. (2012). *PSEP: Fact Sheets: Modern Agriculture: Its Effects on the Environment.* N.p. http://psep.cce.cornell.edu/facts-slides-self/facts/mod-ag-grw85.aspx, (August 11, 2017).

University of California. (1996). *Home Page—UC Statewide IPM Program.* http://ipm.ucanr.edu/, (August 11, 2017).

Unlu K, Ozenirler G, Yurteri C (1999) Nitrogen fertilizer leaching from cropped and irrigated sandy soil in Central Turkey. Eur J Soil Sci 50 (4): 609–620.

US Bureau of Reclamation. (2008). *The Law of the River.* https://www.usbr.gov/lc/region/g1000/lawofrvr.html, (August 11, 2017).

USDA. (2004). "The Healthy Forests Restoration Act." *The Healthy Forests Restoration Act—Safopedia.* http://www.encyclopediaofforestry.org/index.php/The_Healthy_Forests_Restoration_Act, (August 11, 2017).

USDA ERS - Irrigation & Water Use. (2008). "Overview." https://www.ers.usda.gov/topics/farm-practices-management/irrigation-water-use.aspx, (August 11, 2017).

U.S. Department of Energy, "DOE's Early Investment in Shale Gas Technology Producing Results Today."www.netl.doe.gov/publications/press/2011/11008-DOE_Shale_Gas_Research_Producing_R.html, (February 2, 2011).

U.S. Energy Information Administration—EIA—Independent Statistics and Analysis. (2016). *U.S. Energy Information Administration (EIA).* N.p. https://www.eia.gov/analysis/studies/usshalegas/, (August 11, 2017).

US EPA. (1994). "Evaluation of Ecological Impacts from Highway Development." https://www.epa.gov/sites/production/files/2014-08/documents/ecological-impacts-highway-development-pg_0.pdf, (August 11, 2017).

US EPA. (2000). Epa, Ow Us. "US EPA." *Principles of Wetland Restoration Wetlands Protection and Restoration.* N.p. https://www.epa.gov/wetlands/principles-wetland-restoration, (August 11, 2017).

US EPA. (2012). Epa, Ow Us. "US EPA." *Handbook for Developing Watershed Plans to Restore and Protect Our Waters Polluted Runoff: Nonpoint Source Pollution.* N.p. https://www.epa.gov/nps/handbook-developing-watershed-plans-restore-and-protect-our-waters, (August 11, 2017).

US EPA. (2016a). Epa, Ow Us. "US EPA." *Vernal Pools Wetlands Protection and Restoration /.* N.p. https://www.epa.gov/wetlands/vernal-pools, (August 11, 2017).

US EPA. (2016b). *Oil and Gas Extraction Effluent Guidelines Effluent Guidelines .* https://www.epa.gov/eg/oil-and-gas-extraction-effluent-guidelines, (August 11, 2017).

US Fish and Wildlife Service. (1981). *Mitigation Policy.* http://www.fws.gov/policy/501fw2.html.

US Forest Service. (2013). "Forest Service National Strategic Framework for Invasive Species Management." https://www.fs.fed.us/foresthealth/publications/Framework_for_Invasive_Species_FS-1017.pdf, (August 11, 2017).

US Forest Service. (2015). "Bark Beetles Are Decimating Our Forests. That Might Actually Be a Good Thing." *Bark Beetles Are Decimating Our Forests. That Might Actually Be a Good Thing. – Mother Jones.* http://www.motherjones.com/environment/2015/03/bark-pine-beetles-climate-change-diana-six/, (August 11, 2017).

Vengosh, A. (2013). *Salinization and Saline Environments.* Elsevier Science. Vol. 9.

Vidic, R.D., S.L. Brantley, J.M. Vandenbossche, D. Yoxtheimer, and J.D. Abad. (2013). "Impact of Shale Gas Development on Regional Water Quality." *Science* 340, no. 6134, p. 1235009.

Warner, N., Z. Lgourna, L. Bouchaou, S. Boutaleb, T. Tagma, M. Hsaissoune and A. Vengosh. (2013). "Integration of geochemical and isotopic tracers for elucidating water sources and salinization of shallow aquifers in the sub-Saharan Drâa Basin, Morocco." *Applied Geochemistry* 34, pp. 140–151.

Weyerhauser (2017). *Weyerhaeuser: Sustainability.* N.p. https://www.weyerhaeuser.com/sustainability/, (August 11, 2017).

Wilson, J., Y. Wang, and J. VanBriesen. (2013). "Sources of High Total Dissolved Solids to Drinking Water Supply in Southwestern Pennsylvania." *Journal of Environmental Engineering* 140, no. 5.

Wilson, J., and J.M. VanBriesen. (2012). "Oil and Gas Produced Water Management and Surface Drinking Water Sources in Pennsylvania." *Environmental Practice* 14, pp. 288–300.

Witman, S. (2017). "Deforestation Effects as Different as Night and Day." *Deforestation Effects as Different as Night and Day—Eos.* N.p. https://eos.

org/research-spotlights/deforestation-effects-as-different-as-night-and-day, (August 11, 2017).

WRI (2015). From Copenhagen To Cancun: Forests and REDD World Resources Institute. [online] Available at: http://www.wri.org/blog/2010/05/copenhagen-cancun-forests-and-redd [Accessed 20 Sep. 2017].

WWF. (2017). "Environmental Impacts of Farming." *Environmental Impacts of Farming*. http://wwf.panda.org/what_we_do/footprint/agriculture/impacts/.

Younger, P.L., S.A. Banwart, and R.S. Hedin. (2002). *Mine Water: Hydrology, Pollution, Remediation*. Dordrecht, The Netherlands: Kluwer Academic Publishers.

Author Biography

Dr. Alandra Kahl currently teaches engineering design and sustainable systems at the Pennsylvania State University, Greater Allegheny campus. She received her doctorate in environmental engineering from the University of Arizona in 2013, where her dissertation focused on the fate and transport of contaminants of emerging concern in an arid region. Her first book, *Introduction to Environmental Engineering*, was published by Momentum Press in 2015. Dr. Kahl's research interests include engineering of sustainable systems, treatment of emerging contaminants via natural systems and engineering education. She is the author of several books, technical papers, and conference proceedings centered on environmental engineering. Her professional affiliations include the Society of Toxicology and Chemistry, the American Chemical Society, and the American Society for Engineering Education.

INDEX

OTHER TITLES IN OUR ENVIRONMENTAL ENGINEERING COLLECTION

Francis J. Hopcroft, Wentworth Institute of Technology, *Editor*

- *Engineering Economics for Environmental Engineers* by Francis J. Hopcroft
- *Ponds, Lagoons, and Wetlands for Wastewater Management* by Matthew E. Verbyla
- *Environmental Engineering Dictionary of Technical Terms and Phrases: English to Farsi and Farsi to English* by Francis J. Hopcroft and Nima Faraji
- *Environmental Engineering Dictionary of Technical Terms and Phrases: English to Turkish and Turkish to English* by Francis J. Hopcroft and A. Ugur Akinci
- *Environmental Engineering Dictionary of Technical Terms and Phrases: English to Vietnamese and Vietnamese to English* by Francis J. Hopcroft and Minh N. Nguyen
- *Environmental Engineering Dictionary of Technical Terms and Phrases: English to Hungarian and Hungarian to English* by Francis J. Hopcroft and Gergely Sirokman

Momentum Press offers over 30 collections including Aerospace, Biomedical, Civil, Environmental, Nanomaterials, Geotechnical, and many others. We are a leading book publisher in the field of engineering, mathematics, health, and applied sciences.

Momentum Press is actively seeking collection editors as well as authors. For more information about becoming an MP author or collection editor, please visit http://www.momentumpress.net/contact

Announcing Digital Content Crafted by Librarians

Concise e-books business students need for classroom and research

Momentum Press offers digital content as authoritative treatments of advanced engineering topics by leaders in their field. Hosted on ebrary, MP provides practitioners, researchers, faculty, and students in engineering, science, and industry with innovative electronic content in sensors and controls engineering, advanced energy engineering, manufacturing, and materials science.

Momentum Press offers library-friendly terms:
- *perpetual access for a one-time fee*
- *no subscriptions or access fees required*
- *unlimited concurrent usage permitted*
- *downloadable PDFs provided*
- *free MARC records included*
- *free trials*

The **Momentum Press** digital library is very affordable, with no obligation to buy in future years.

For more information, please visit **www.momentumpress.net/library** or to set up a trial in the US, please contact **mpsales@globalepress.com**.

CPSIA information can be obtained
at www.ICGtesting.com
Printed in the USA
FFHW011214120619
52960349-58558FF